Contents

Glossary

As you work through this book, add the meanings to the words.

Aerial photograph ______________________________

Anticyclone ______________________________

Atmosphere ______________________________

Axis ______________________________

Bar graph ______________________________

Best-fit line ______________________________

Birth rate ______________________________

Cartoon ______________________________

Choropleth ______________________________

Climograph ______________________________

Compass rose ______________________________

Contemporary ______________________________

Continuum ______________________________

Contour line ______________________________

Cross-section ______________________________

Cultural ______________________________

Cycle ______________________________

Data ______________________________

Death rate ______________________________

Decrease ______________________________

Degree ______________________________

Demographic ______________________________

Density ______________________________

Depression (of relief) ______________________________

Depression (of weather) ______________________________

Direction ______________________________

Dispersed ______________________________

Easting ______________________________

Equator ______________________________

Fact ______________________________

Feature ______________________________

Feedback ______________________________

Flow chart ______________________________

Front (of weather) ______________________________

GEED ______________________________

Generalisation ______________________________

Geographical ______________________________

Global ______________________________

Hemisphere ______________________________

Horizontal ______________________________

Increase ______________________________

Inference ______________________________

Inputs ______________________________

 ISBN: 9780170368131

Irregular

Isobar

Issue

Key

Latitude

Line graph

Linear

Location

Longitude

Map

Mean

Model

Natural

Negative

Northing

Nuclear

Oblique

Observation

Opinion

Outputs

Pattern

Percentage bar graph

Pie graph

Population pyramid

Positive

Precis

Pressure reading

Projected

Questionnaire

Random

Regular

Relationship

Representative fraction

Resource

Scale

Spatial

Star diagram

Survey

Symbol

System

Topography

Transition

Trapezium

Tropical cyclone

Values

Variable

Venn diagram

Vertical

Visual summary

Weather map

ISBN: 9780170368131

02

Percentages

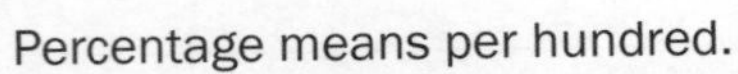

Percentage means per hundred.

It comes from the Latin *per centum* meaning by the hundred.

Percentage symbol = %

For example, there are 100 ticks in this box.

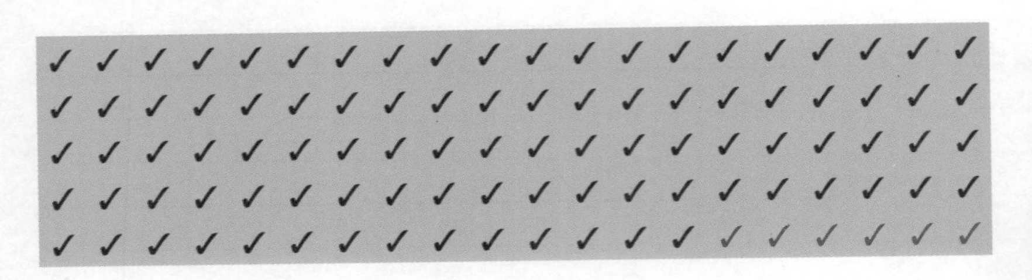

Say this stands for the whole population of Hong Kong. The black ticks stand for Chinese people; 94 ticks are black. This means 94 percent of people in Hong Kong are Chinese. The red ticks stand for non-Chinese; 6 ticks are red. This means 6 percent of people in Hong Kong are non-Chinese.

Percentages can be shown as fractions and vice versa.

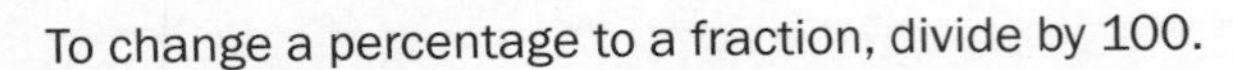

To change a percentage to a fraction, divide by 100.

For example: $20\% = \frac{20}{100} = \frac{1}{5}$

20% 20% 20% 20% 20% =100%

To change a fraction to a percentage, multiply by 100.

For example: $\frac{1}{5} \times \frac{100}{1} = \frac{100}{5} = 20$, so $\frac{1}{5} = 20\%$

Why it's useful to be able to work out percentages

- It's a skill you can use in all your subjects and for all your life.
- It helps you with other skills such as drawing graphs.
- Percentages are an easy way to show data (facts, statistics).

How to work out a percentage

For example: New Zealand's population figures in 2014

0–14 years old	911,300
15–64 years old	2,948,200
65+ years old	650,400
TOTAL	4,509,900

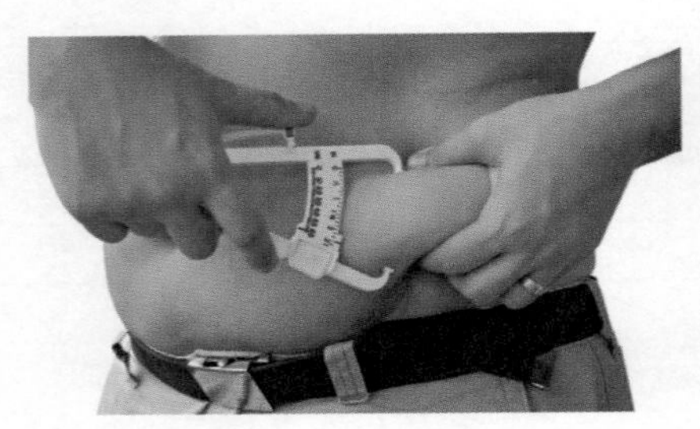

 ISBN: 9780170368131

Use the following formula to put the population figures into percentages.

$$\frac{100}{1} \times \frac{\text{number}}{\text{total number}} = \text{percentage}$$

0–14 years old

$$\frac{100}{1} \times \frac{911,300}{4,509,900} = 20.21\%$$

15–64 years old

$$\frac{100}{1} \times \frac{2,948,200}{4,509,900} = 65.37\%$$

65+ years old

$$\frac{100}{1} \times \frac{650,400}{4,509,900} = 14.42\%$$

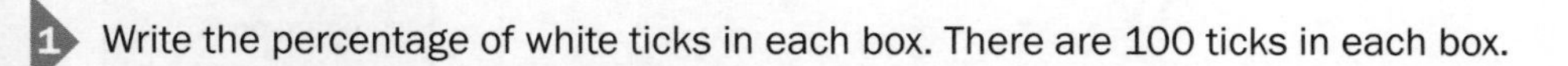

1 Write the percentage of white ticks in each box. There are 100 ticks in each box.

a ______________________ **b** ______________________

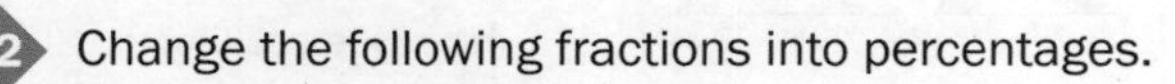

2 Change the following fractions into percentages.

a $\frac{1}{4}$ ______________________ **b** $\frac{2}{3}$ ______________________

c $\frac{1}{5}$ ______________________ **d** $\frac{3}{5}$ ______________________

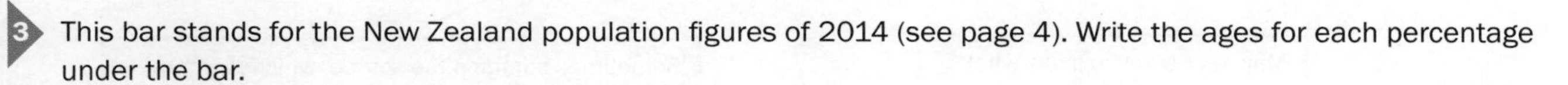

3 This bar stands for the New Zealand population figures of 2014 (see page 4). Write the ages for each percentage under the bar.

a	b	c
14.42%	20.21%	65.37%

a ______________________

b ______________________

c ______________________

Total % ______________________

4 Work out the following percentages for the population of Hong Kong. Hong Kong's total population was 7,210,505.

	Age	Number	% of population
a	0–14	1,278,566	
b	15–64	5,156,991	
c	65+	774,948	

ISBN: 9780170368131

03

Column graphs

Example of a column graph

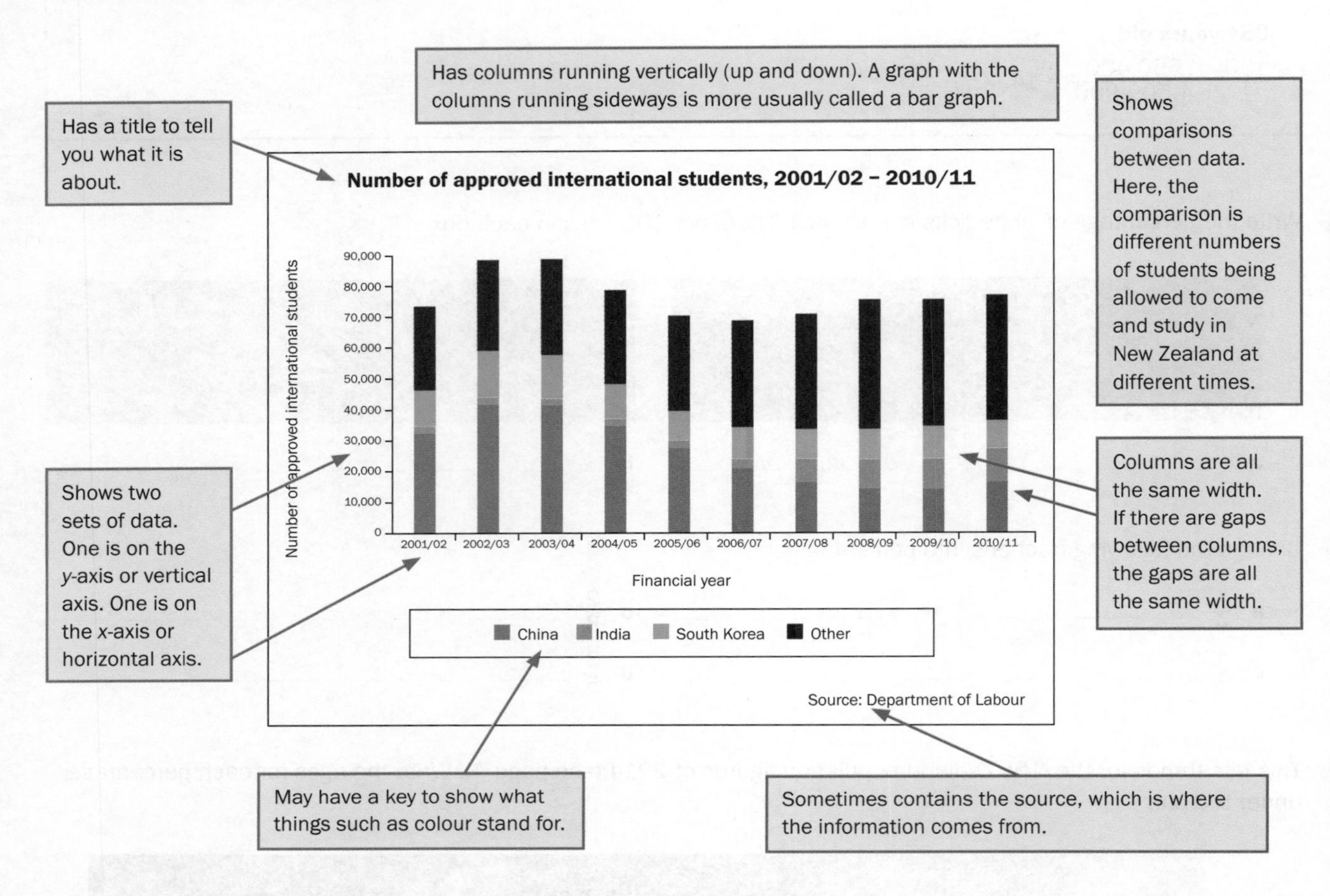

1 Fill out the following about the graph above.

a The graph's source is ____________________

b The vertical axis is also called the ____________________ axis.

c The number of people in thousands is shown on the ____________________ axis.

d The horizontal axis is also called the ____________________ axis.

e The dates are shown on the ____________________ axis.

f The width of each column is ____________________ mm.

g The width of the gap between each column is ____________________ mm.

ISBN: 9780170368131

2 Fill out the following about the graph below.

a There is a total of ______________________ columns on the graph.

b The qualification data is shown on the ______________________ axis.

c The income data is shown on the ______________________ axis.

d The highest figure for income is for ______________________ qualification.

e The income for NCEA 1 qualification is ______________________

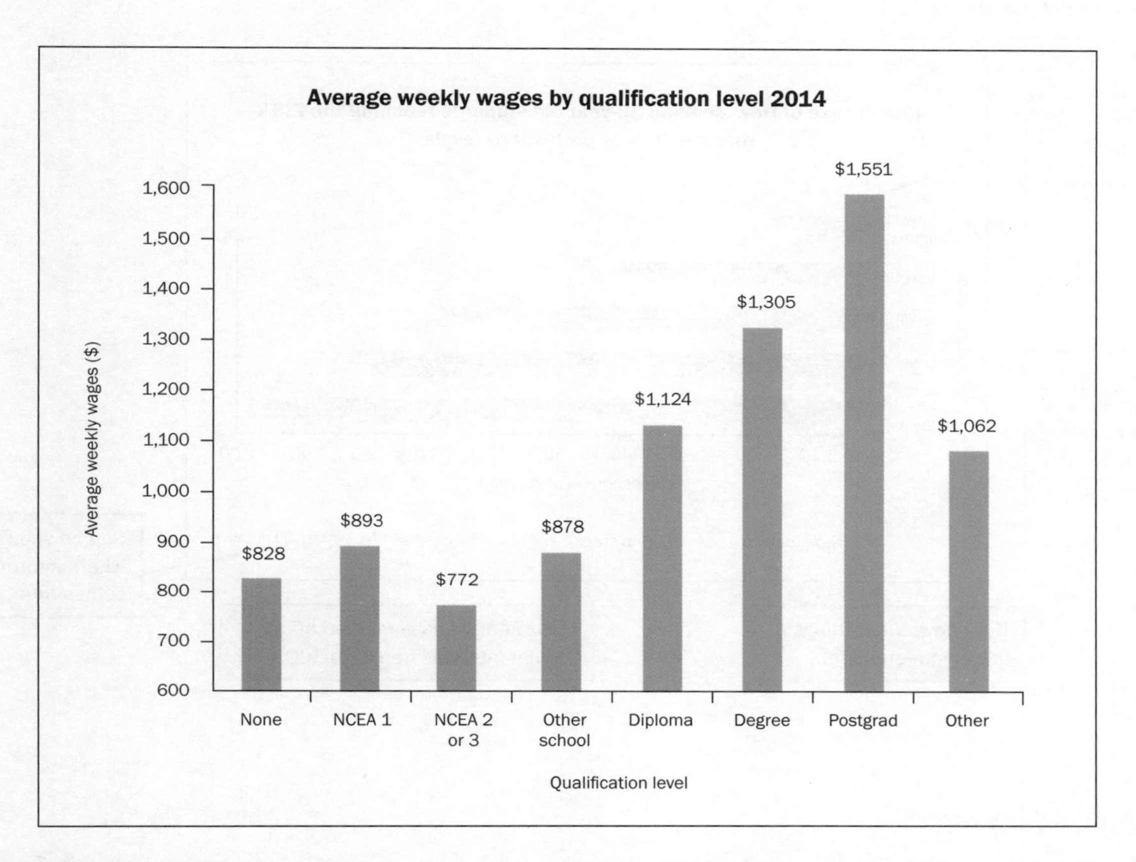

3 Put the following figures in the right places on the graph below, which shows some projected (estimated for the future) figures.

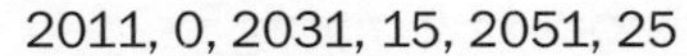
2011, 0, 2031, 15, 2051, 25

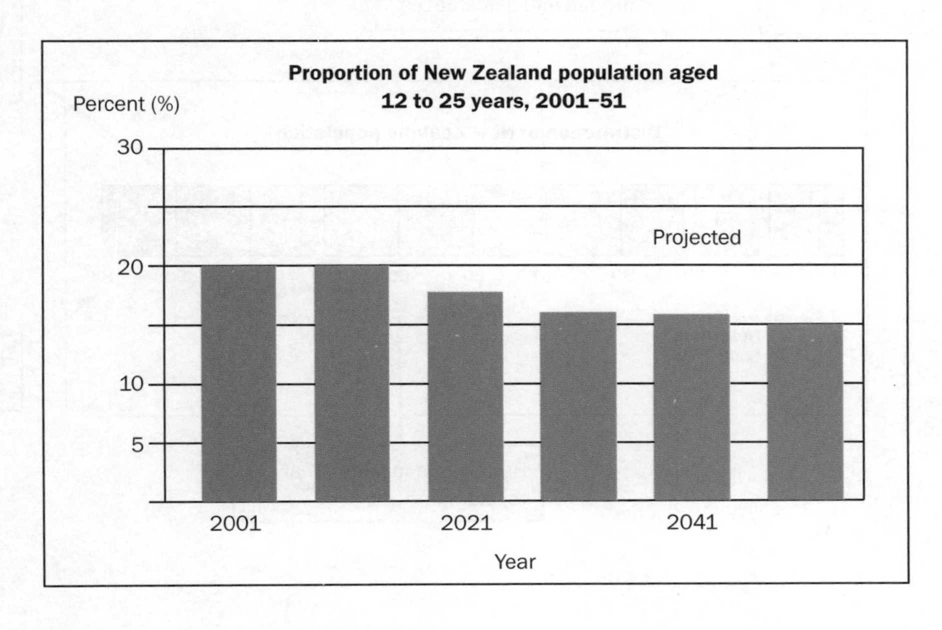

ISBN: 9780170368131

04

Bar graphs

Proficiency = skill

PISA = Programme for International Student Assessment

OECD = Organisation for Economic Co-operation and Development

Example of a bar graph

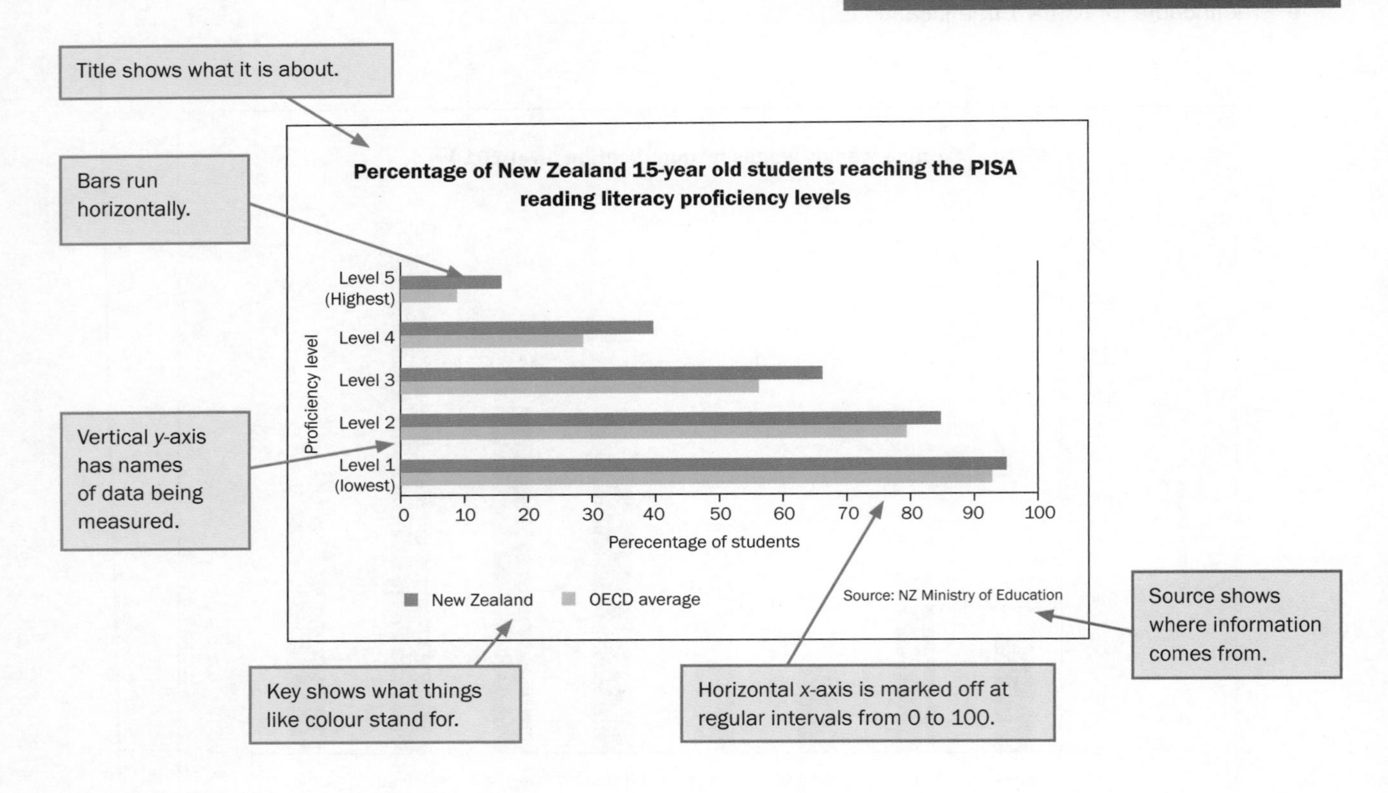

Percentage bar graph

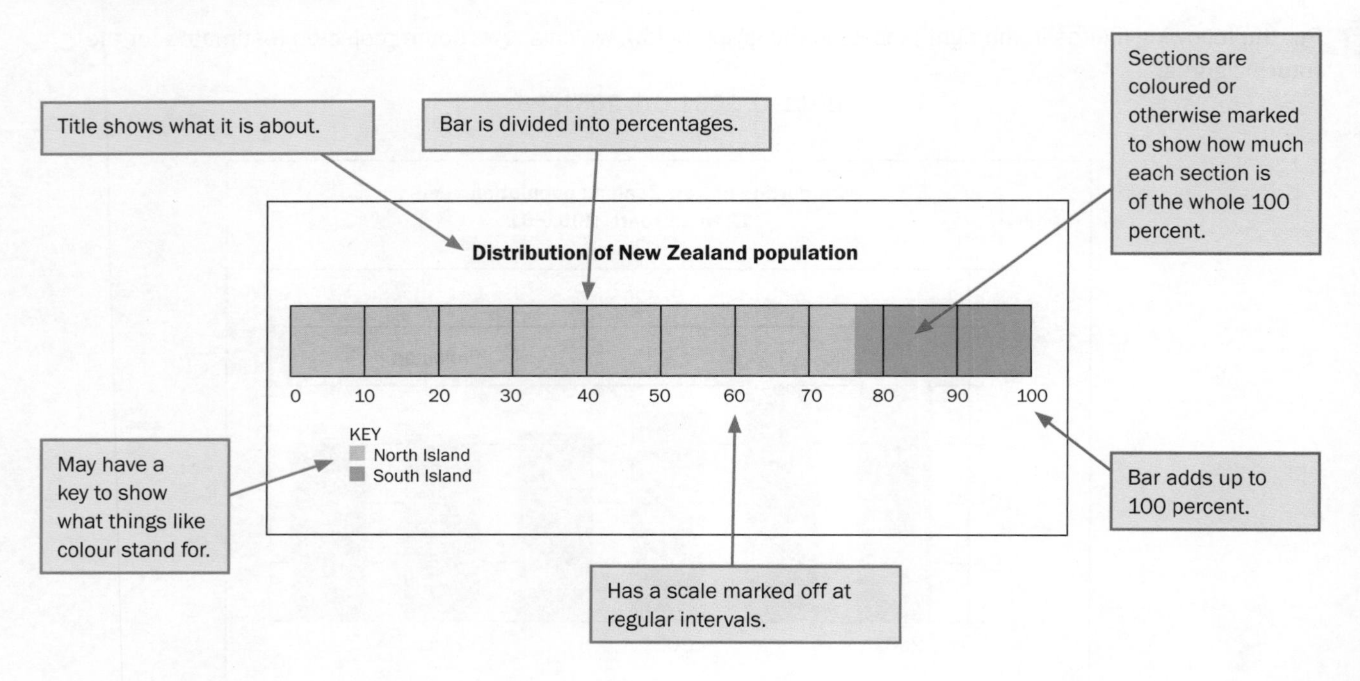

ISBN: 9780170368131

1 a Write the figures for the percent scales of the two graphs below.

b Use two colours to show that Maori were 14.9% of the total population in 2014 and 1% of the total population in 1901.

c Finish the key by adding the right colours.

Graph 1 Percentage of Maori in New Zealand population 2014

0 100 percent

KEY:

☐ Maori

☐ Non-Maori

Graph 2 Percentage of Maori in New Zealand population 1901

0 100 percent

2 Add the following to the graph below.

60, Source:, disaster, clothes, Percent, Statistics New Zealand, Water, 20, Torch

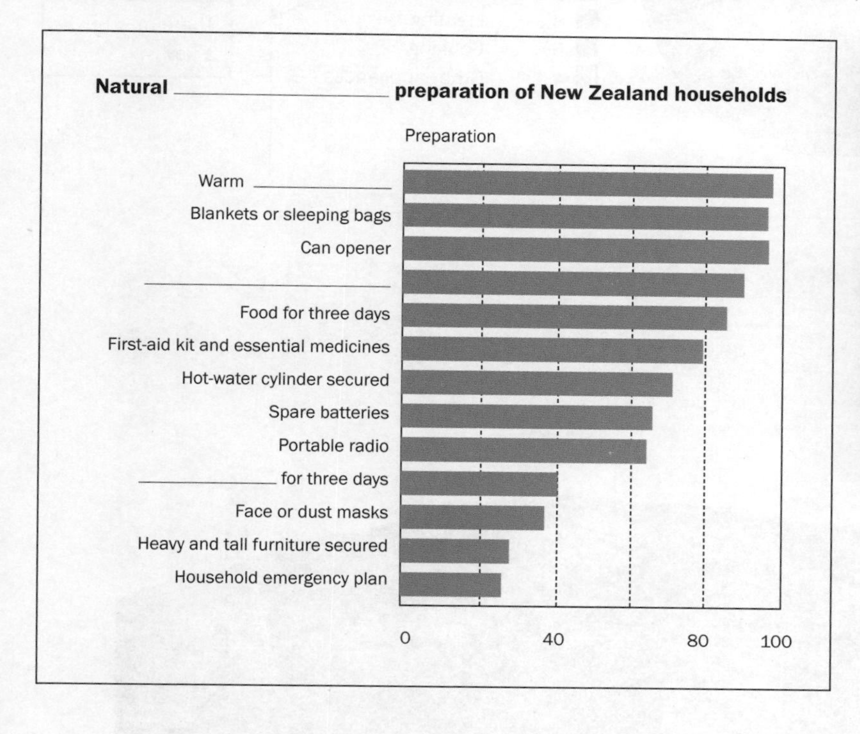

05 Pie graphs

A pie graph is shaped like the old-fashioned pie – it's circular.
Pie graphs are good for showing percentages.

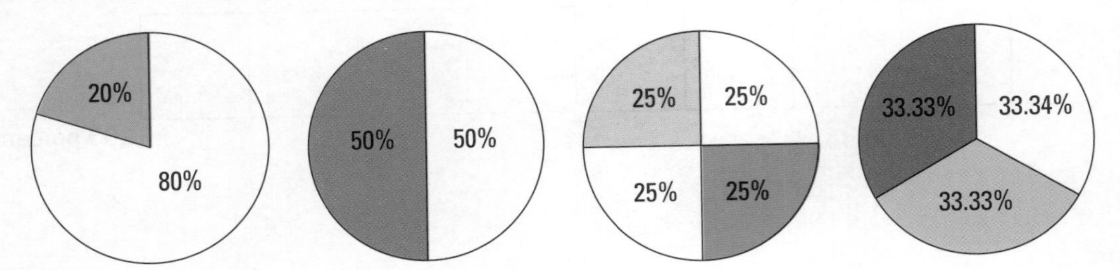

Percentages of the pie graph must add up to 100.

Example of a pie graph

Title tells you what the graph is about.

If the percentages have not been written on the graph, you can work them out by using your eyes to estimate them or you can measure the angle of a slice/sector with a protractor and divide the number of degrees by 3.6, as there are 360 degrees in a circle.

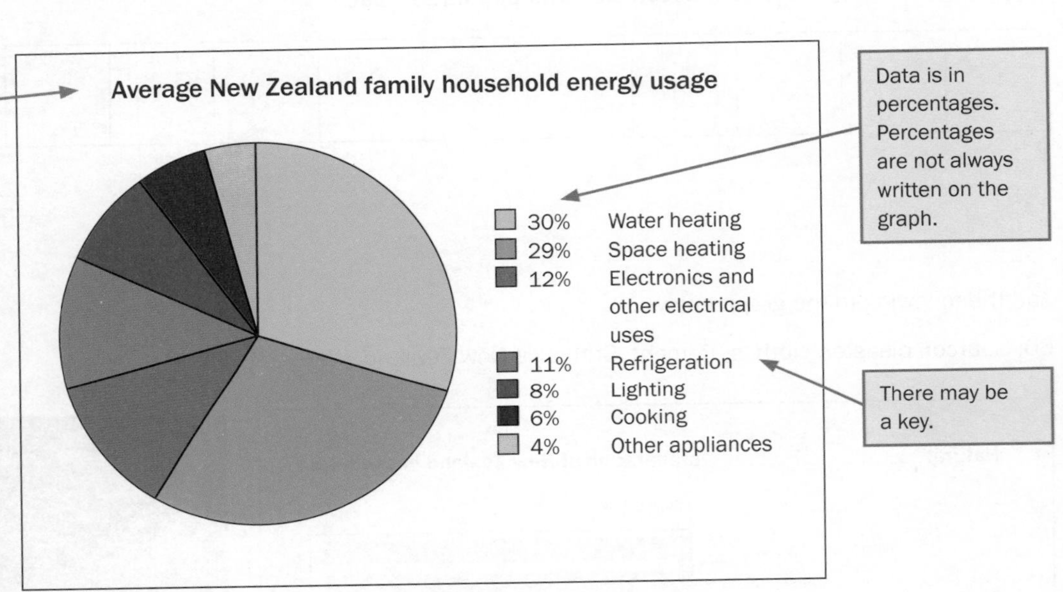

ISBN: 9780170368131

1 Use different colours on the following pie graphs and add your colours to the keys.

a Ethnic percentages in Mexico

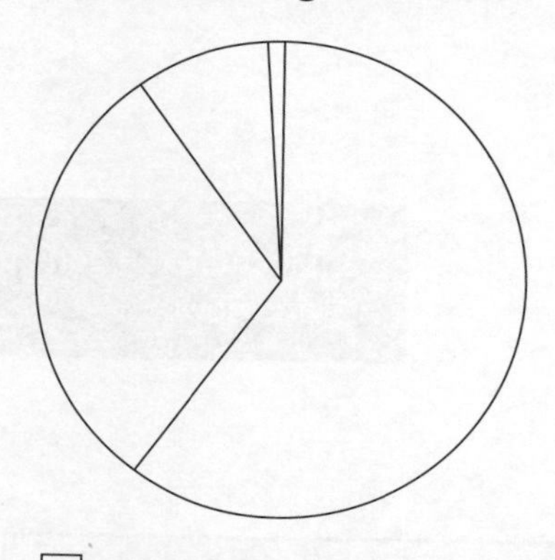

- ☐ Mestizo (Amerindian-Spanish) 60.00
- ☐ Amerindian 30.00
- ☐ White 9.00
- ☐ Other 1.00

b Ethnic percentages in Singapore

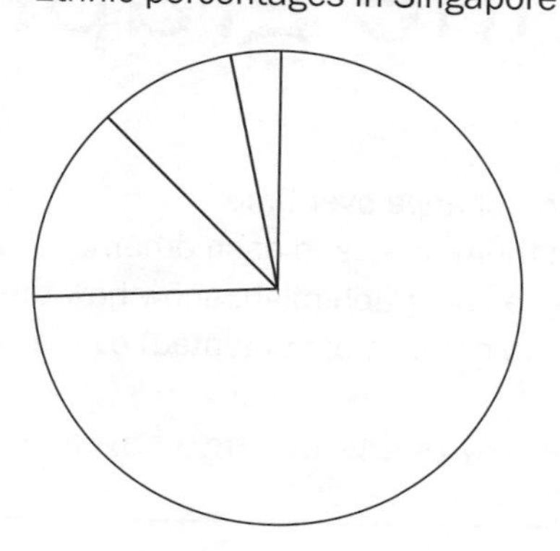

- ☐ Chinese 74.20
- ☐ Malay 13.30
- ☐ Indian 9.20
- ☐ Other 3.30

c Ethnic percentages in New Zealand

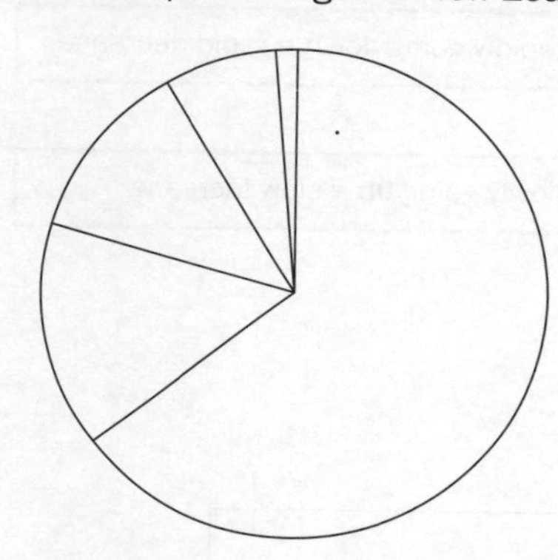

- ☐ NZ European 64.70
- ☐ Maori 14.90
- ☐ Asian 11.80
- ☐ Pacific Islanders 7.40
- ☐ Other 1.20

2 Put the data from the brackets into the right boxes on the graphs.

a Ethnic percentages in Australia

(White 92%, Asian 7%, Aboriginal and other 1%)

b Ethnic percentages in South Africa

(Black African 79.2, White 8.9, Coloured 8.9, Indian/Asian 2.5, Other 0.5)

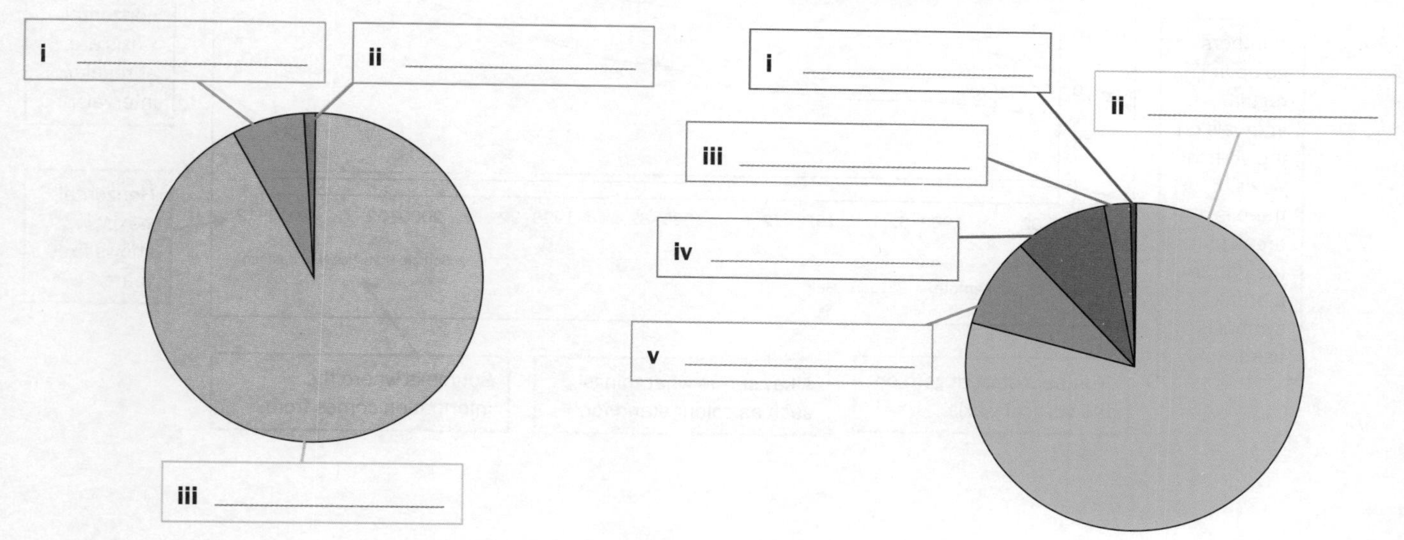

ISBN: 9780170368131

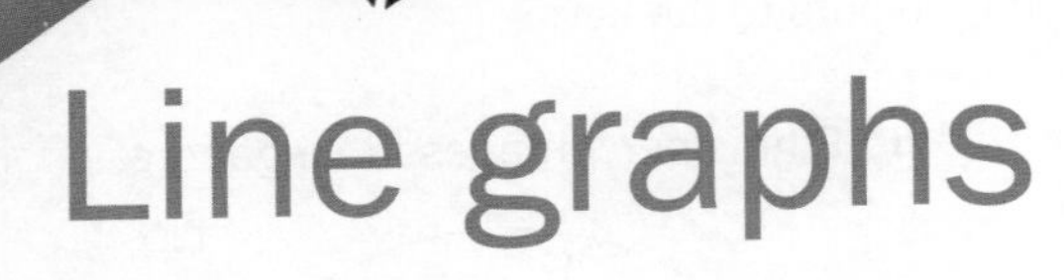

06 Line graphs

Line graphs show change over time.

The data on them relates to each other.

For example, a line graph might show how the population of a place has changed (gone up or down or fluctuated) over a certain number of years.

Fluctuated = gone up and down.

What the lines showing change can look like on a line graph

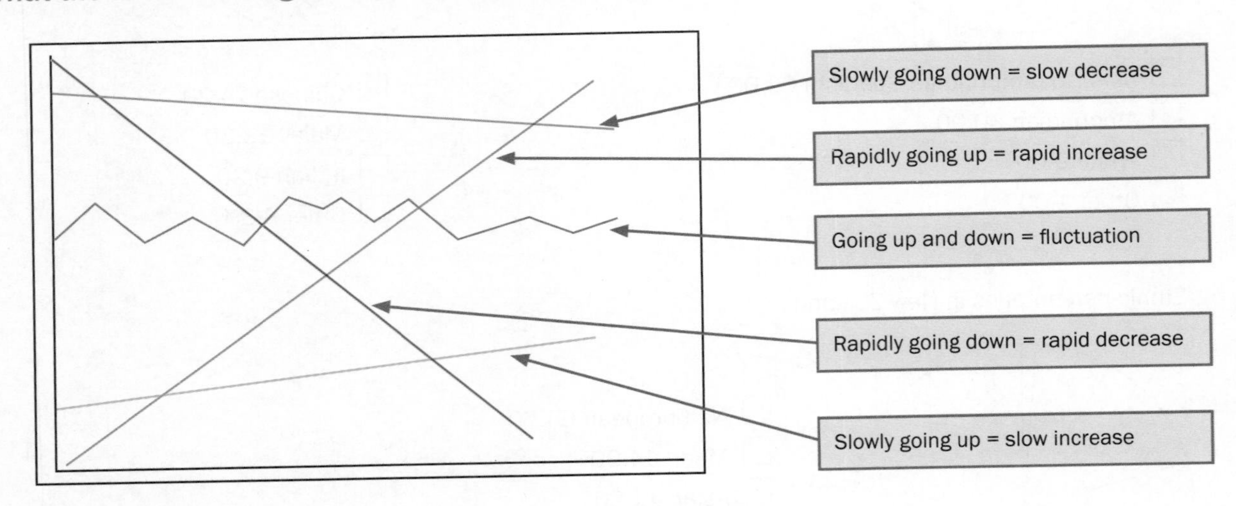

Example of a line graph

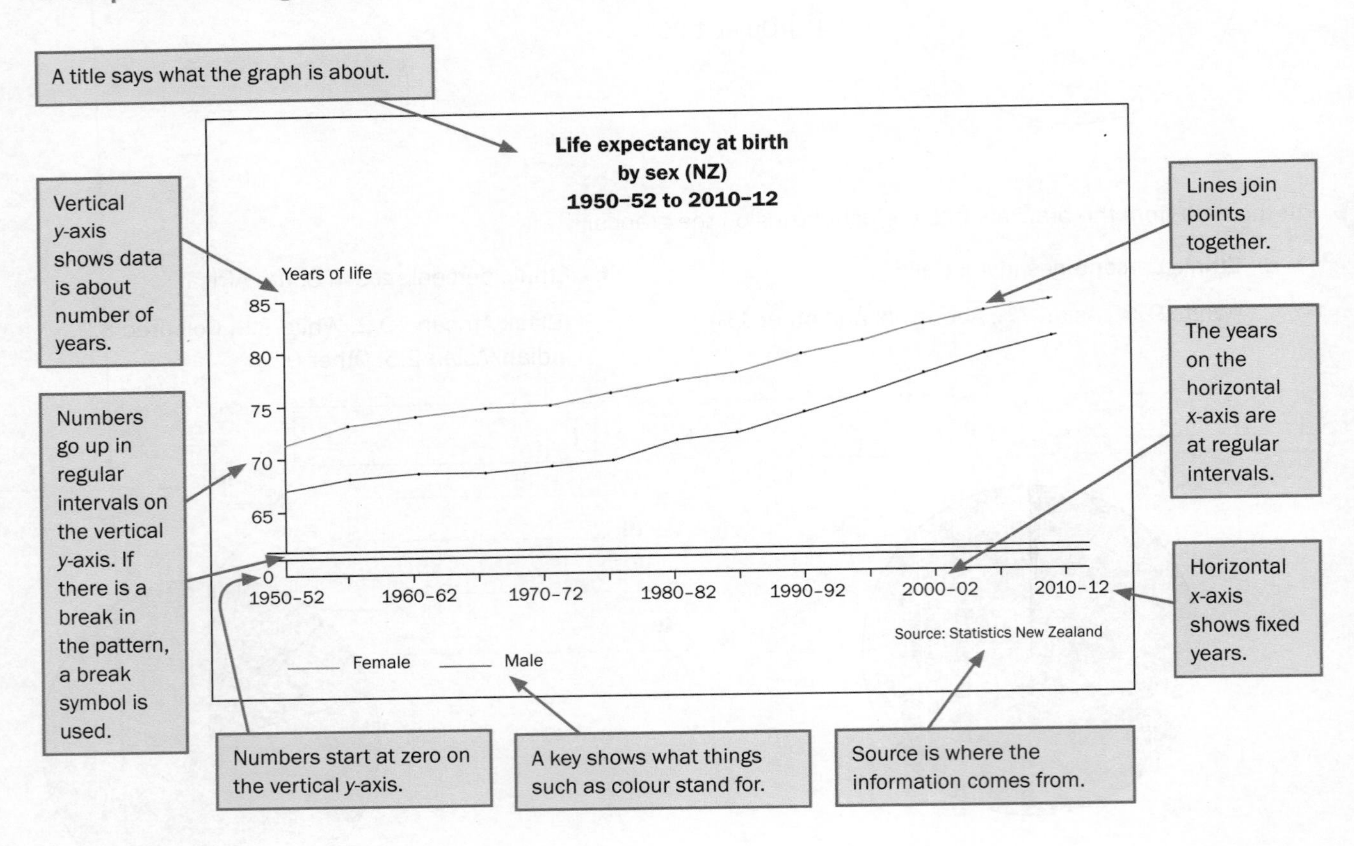

ISBN: 9780170368131

1 In the boxes, write the type of change shown by the graph lines.

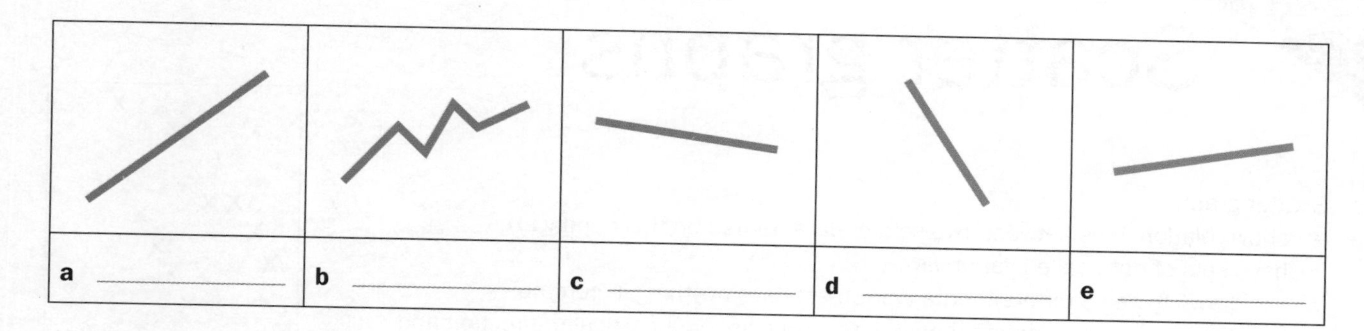

a ______	b ______	c ______	d ______	e ______

2 Join the points on the graph together.

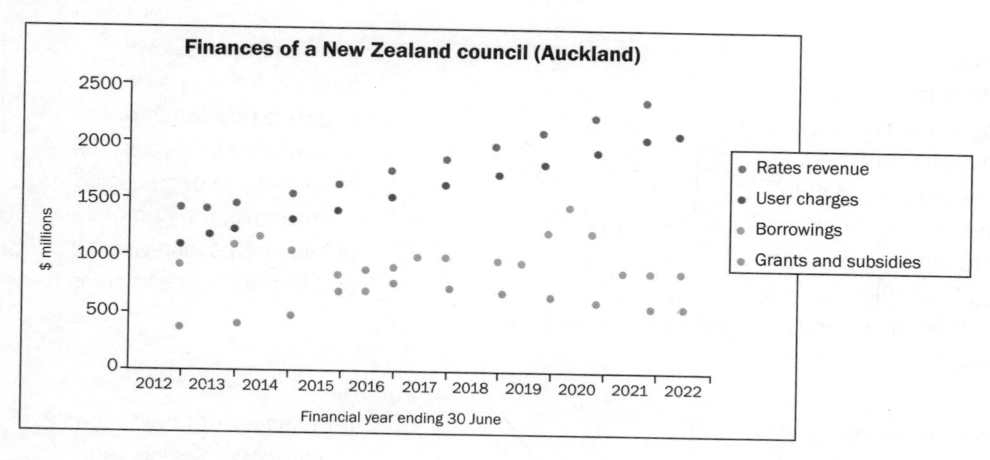

3 Check out the graph and write in the answers about it.

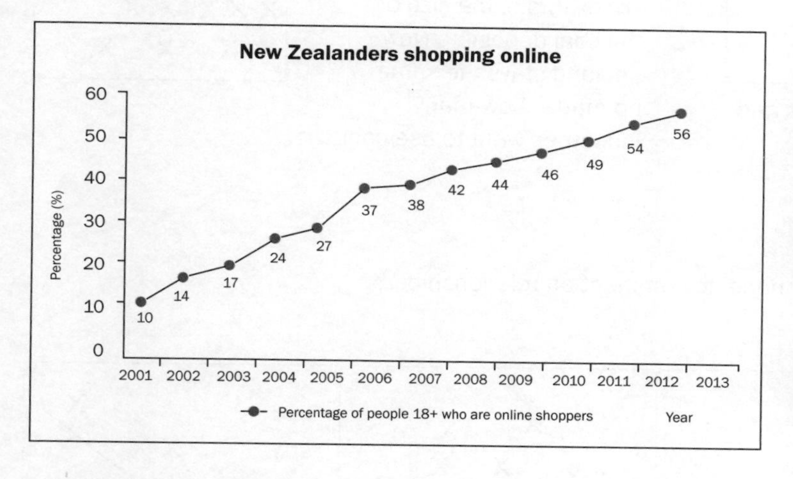

a The graph's title is ______

b If the *y*-axis had figures as high as they could go, the last figure would be ______

c The key shows ______

d The change shown is a ______

e There is no ═ symbol on the graph because ______

f A reason to question the accuracy of the graph is that ______

ISBN: 9780170368131

07

Scatter graphs

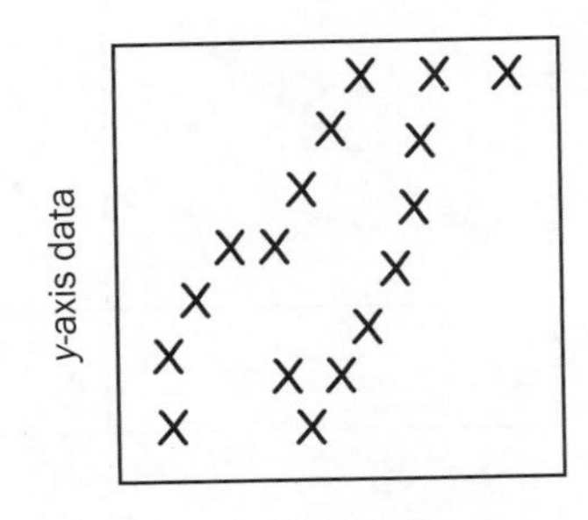

Scatter graphs:

- show relationships between two sets of data named on the *x* and *y* axes
- have sets of data called variables
- are drawn to see if different data (variables) have anything in common
- may have a scatter of points that look to form a line in a particular direction and so it is safe to think there is a relationship between them.

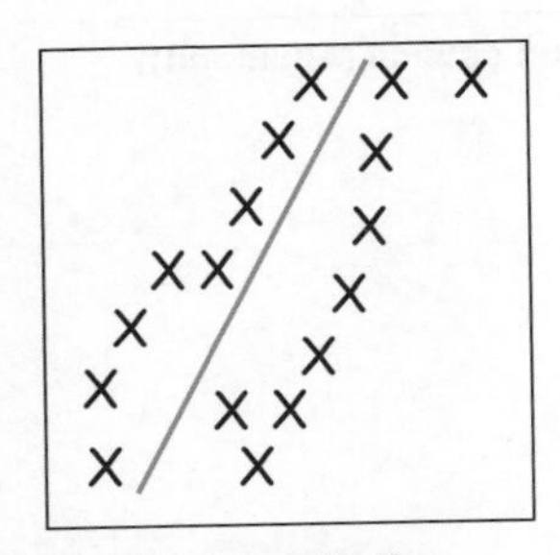

To check for a relationship, put a ruler along the scatter points until about half the points are on one side of the ruler and half the points are on the other side of the ruler. Draw a line along the ruler. This is called the **best fit line** or **trend line**.

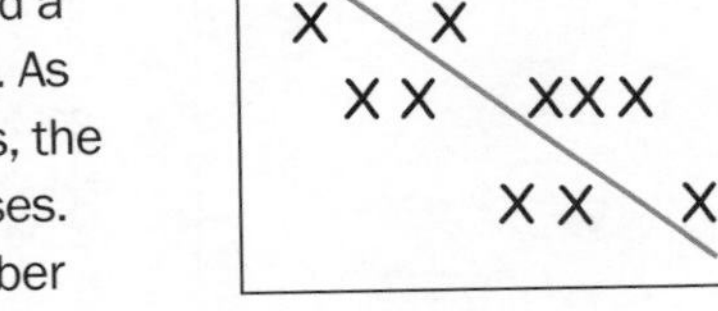

If the line slopes down from the left, it is called a **negative relationship**. As one variable increases, the other variable decreases. For example, the number of bad storms increases and the number of apples harvested decreases.

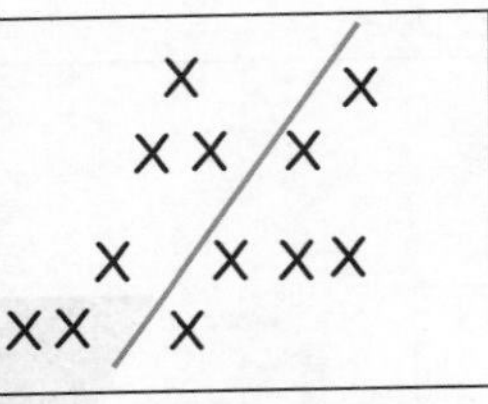

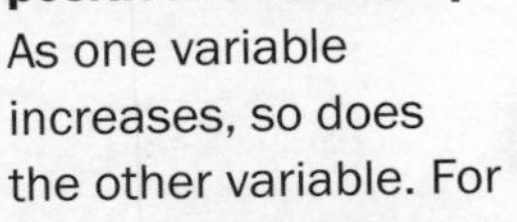

If the line slopes up to the right, it is called a **positive relationship**. As one variable increases, so does the other variable. For example, the population of your town increases and so the number of new houses increases.

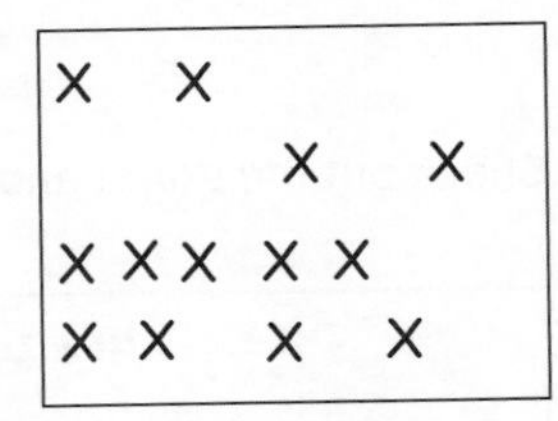

If the scatter of points on the graph shows no line, there is no relationship. For example, the size of the coal deposit in New Zealand stays the same no matter how many industries want to use coal.

1 In the boxes, put either 'positive', 'negative' or 'none' to comment on relationships.

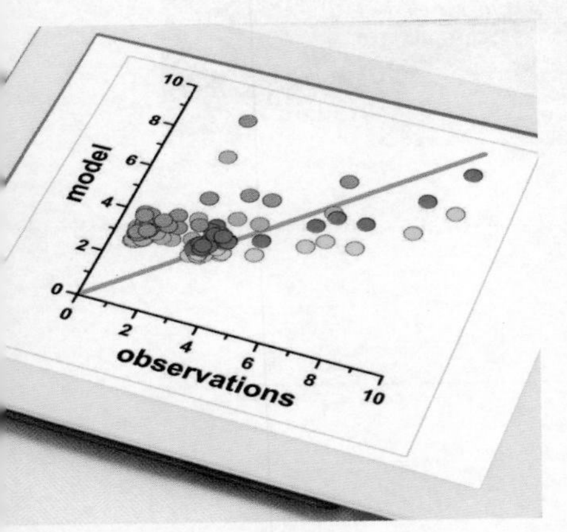

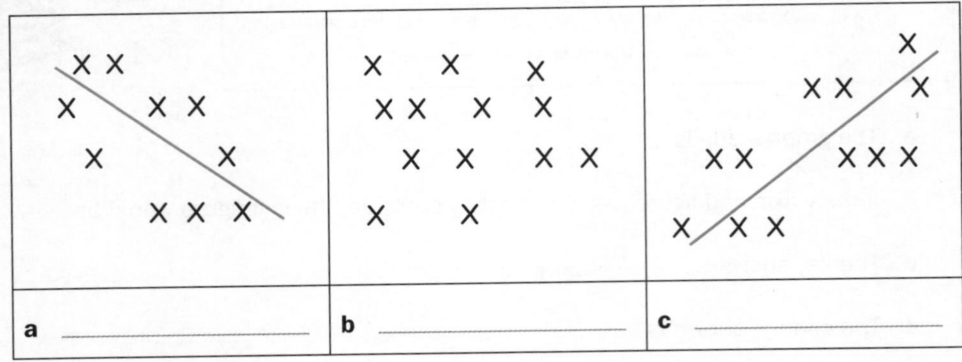

ISBN: 9780170368131

2 Check out the graph and fill in the gaps in the following sentences.

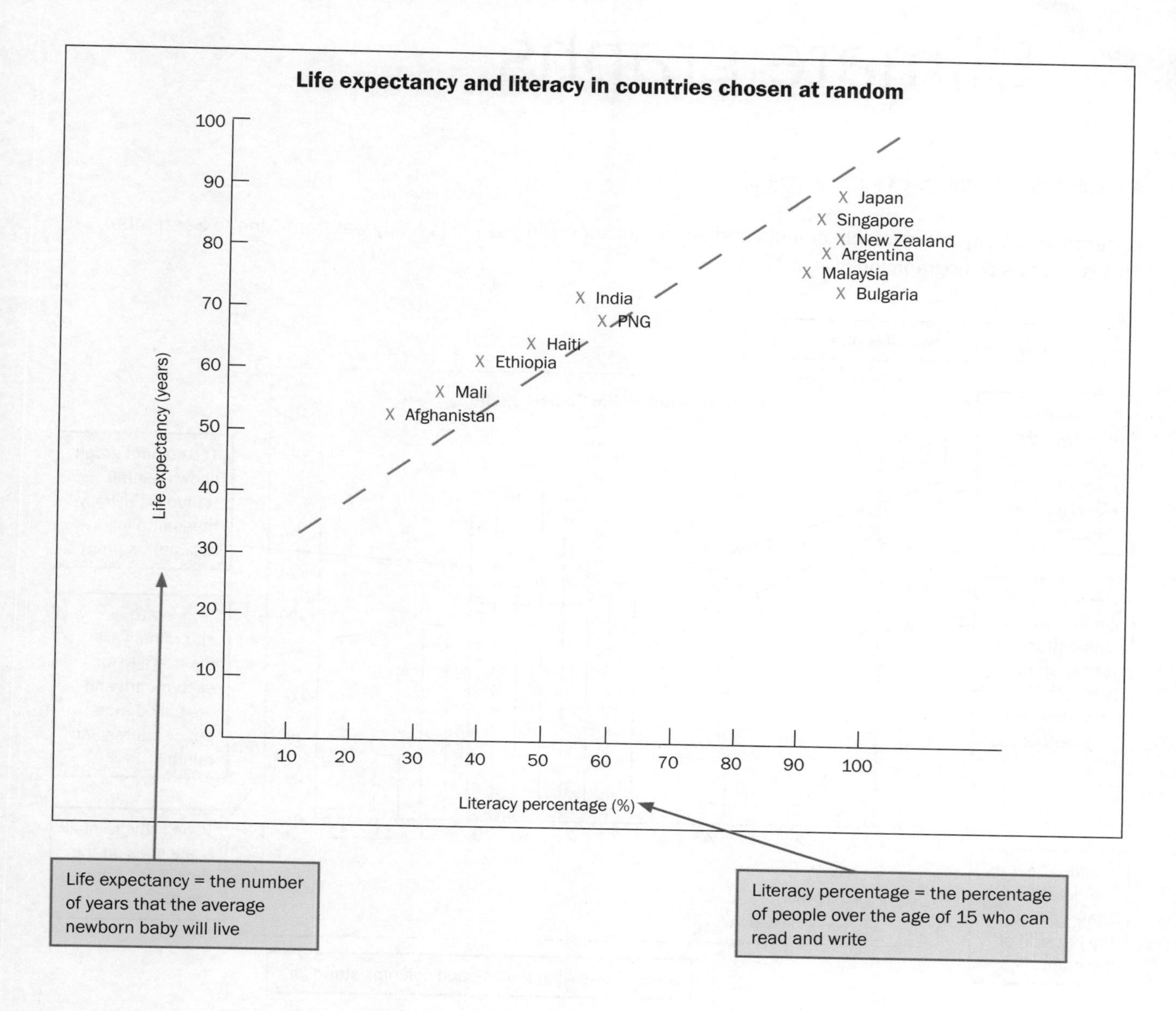

a The title of the graph is ______________________________

b The data along the vertical *y*-axis is ______________________________

c The data along the horizontal *x*-axis is ______________________________

d The name of the line on the graph is ______________________________

e The number of countries on each side of the line is ______________________________

f The line slopes up to the ______________________________

g The type of relationship shown is ______________________________

h As life expectancy increases, literacy ______________________________

ISBN: 9780170368131

08

Climate graphs

A graph about climate is called a climograph.

Hyderabad is a city in the middle of India and has a monsoon climate — it is really wet from June to September. This is what its climograph looks like.

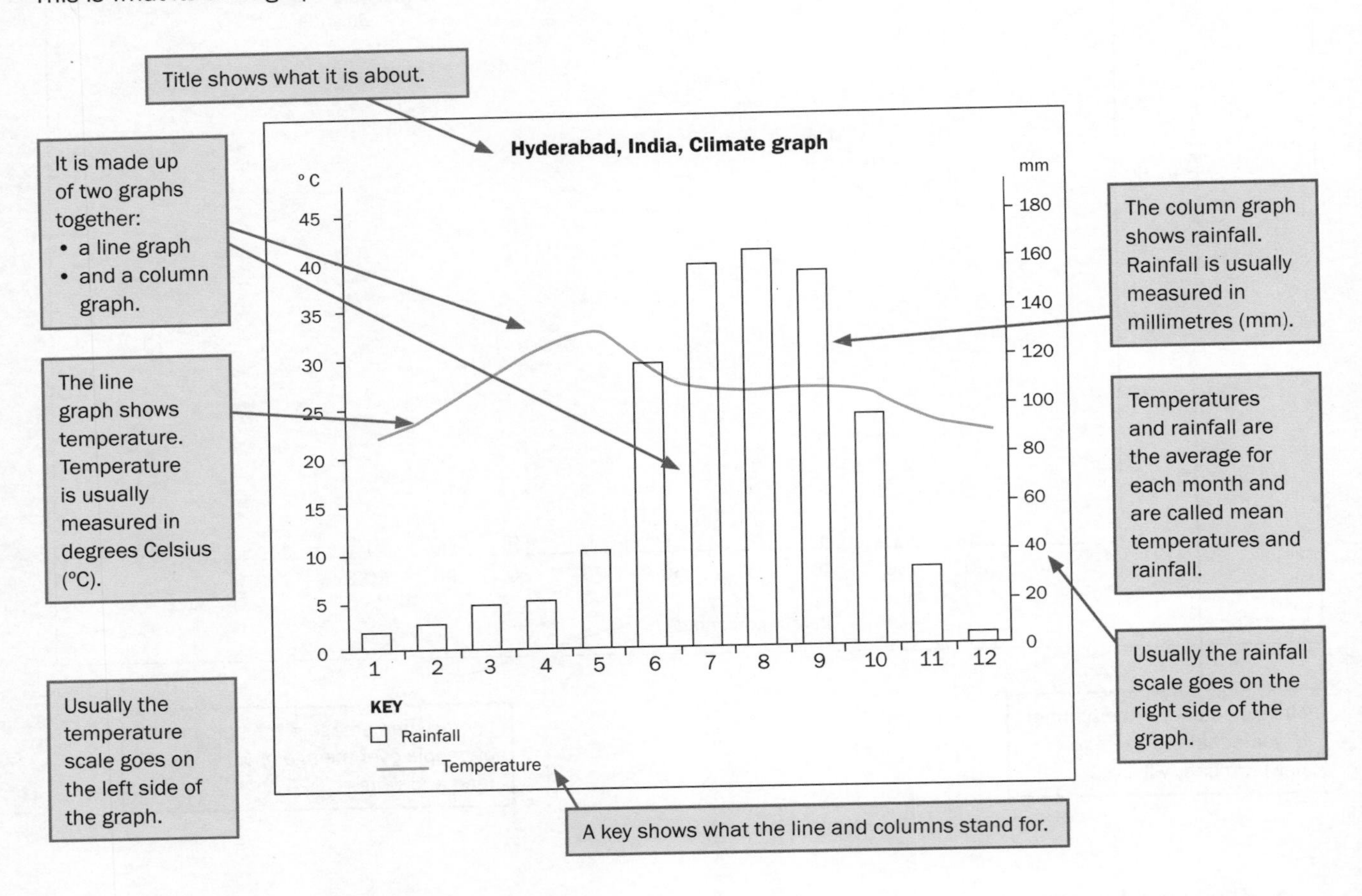

1 Colour the columns on the climograph and then fill in the following data.

a The line graph shows ________________

b The column graph shows ________________

c The numbers on the horizontal *x*-axis stand for ________________

d The measurement used to show rainfall is ________________

e The measurement used to show temperature is ________________

f The highest temperature is ________ in the month of ________

g The lowest temperature is ________ in the month of ________

h The highest rainfall is ________ in the month of ________

i The lowest rainfall is ________ in the month of ________

ISBN: 9780170368131

2 Use the climate data for Beijing, the capital of China, to fill out the climate graph for Beijing. Add colour.

Month	J	F	M	A	M	J	J	A	S	O	N	D
Temperature (°C)	-5	-2	5	14	20	25	26	25	20	13	4	3
Rainfall (mm)	4	5	8	17	35	78	243	141	58	16	11	3

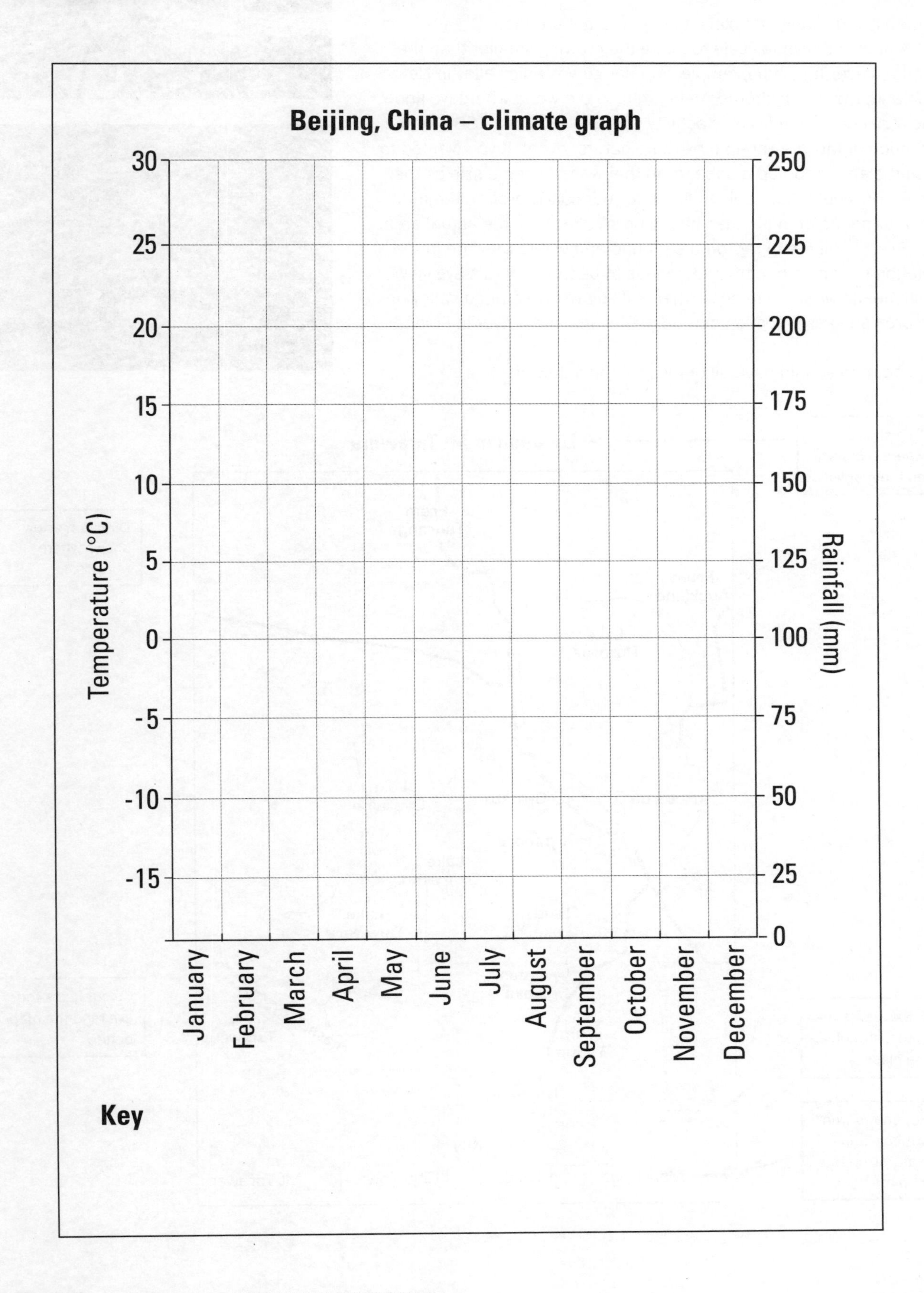

What a map is

A map is a drawing of a place to help you get around.

A map is drawn to scale to make the drawing smaller than the actual place. Take, for example, Waimangu Volcanic Valley in New Zealand, the only hydrothermal system in the world where we know the exact day when surface activity started — 10 June 1886, at the eruption of Mt Tarawera, an extreme natural event. If you wanted to visit the area and had a paper map that was the same size as the area, you wouldn't be able to fit it into your car. Therefore the map on your phone or in an atlas has to be smaller than the actual area.

To show all the things on the ground and where they are in relation to each other, the map needs to be from a bird's-eye view. This means what a bird sees when it flies over Waimangu Valley or an area such as a section of road in Chicago and a river in Poland.

A good map should have all or most of the following.

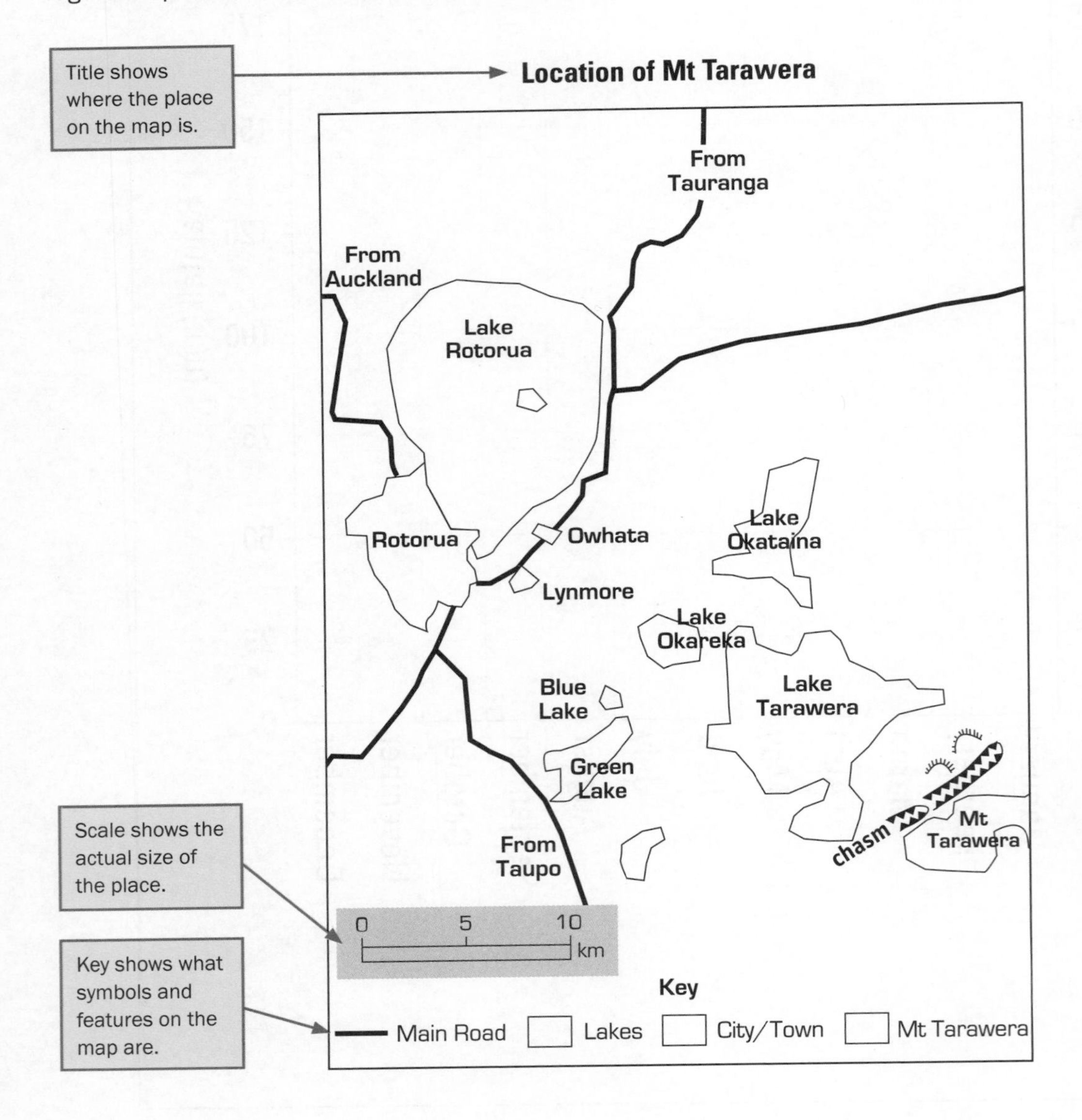

 ISBN: 9780170368131

1 a Write down four things the map on page 18 has that makes it a good map.

i ______________________________

ii ______________________________

iii ______________________________

iv ______________________________

b Write down two things the map could also have had.

i ______________________________

ii ______________________________

2 Colour the map as follows: roads = red, mountain's chasm = brown, lakes = blue, city/town areas = yellow.

3 On this map, add:

a your own colours
b the key, to include your colours
c the frame
d the direction arrow
e a better title
f the following missing information:

21 July 2013, Cook Strait, Magnitude 6.5

17 June 1929, Murchison, Magnitude 7.8

Earthquakes

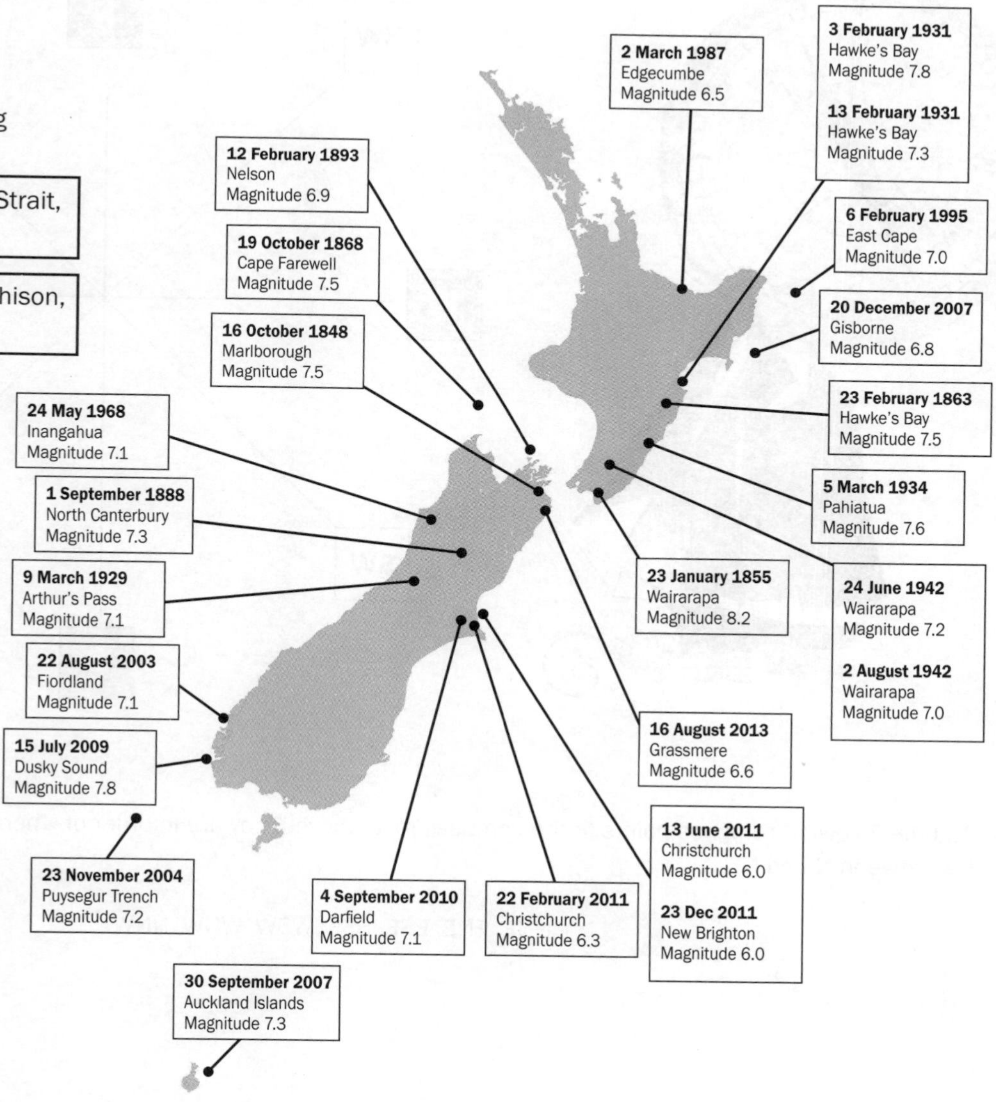

2 March 1987 Edgecumbe Magnitude 6.5

3 February 1931 Hawke's Bay Magnitude 7.8

13 February 1931 Hawke's Bay Magnitude 7.3

12 February 1893 Nelson Magnitude 6.9

19 October 1868 Cape Farewell Magnitude 7.5

6 February 1995 East Cape Magnitude 7.0

16 October 1848 Marlborough Magnitude 7.5

20 December 2007 Gisborne Magnitude 6.8

24 May 1968 Inangahua Magnitude 7.1

23 February 1863 Hawke's Bay Magnitude 7.5

1 September 1888 North Canterbury Magnitude 7.3

5 March 1934 Pahiatua Magnitude 7.6

9 March 1929 Arthur's Pass Magnitude 7.1

23 January 1855 Wairarapa Magnitude 8.2

24 June 1942 Wairarapa Magnitude 7.2

2 August 1942 Wairarapa Magnitude 7.0

22 August 2003 Fiordland Magnitude 7.1

16 August 2013 Grassmere Magnitude 6.6

15 July 2009 Dusky Sound Magnitude 7.8

13 June 2011 Christchurch Magnitude 6.0

23 Dec 2011 New Brighton Magnitude 6.0

23 November 2004 Puysegur Trench Magnitude 7.2

4 September 2010 Darfield Magnitude 7.1

22 February 2011 Christchurch Magnitude 6.3

30 September 2007 Auckland Islands Magnitude 7.3

ISBN: 9780170368131

10

Directions

Direction = the point towards which you face or move

Compass = an instrument used to find directions. It has a magnetised needle that points north.

Compass rose = the pattern on a map that shows the direction of north. Its name comes from the way the points of a compass look like the arrangement of petals of a rose flower.

The main (cardinal) compass points are N (north), E (east), S (south), W (west).

The news you view online, in the paper or on TV, comes from all parts of the world – north, east, west, south.

This compass rose has eight points.

It could be further divided into 16 points to make it even more accurate.

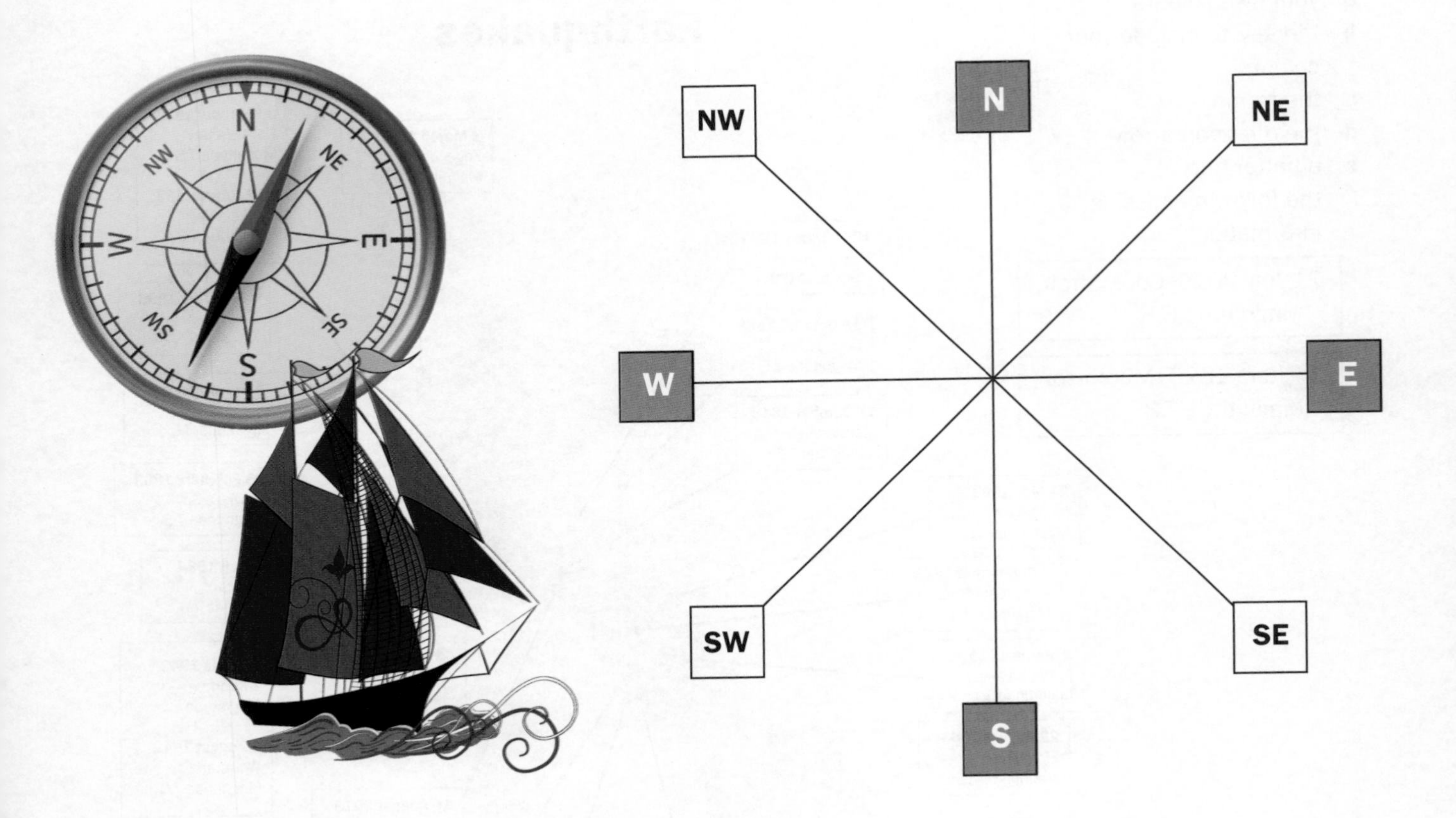

1 Add the following compass points to the compass rose above. They are in order of where they fit; for example, NNE fits between N and NE.

NNE, ENE, ESE, SSE, WSW, WNW, NNW

ISBN: 9780170368131

2 Write in each box the direction the arrow is pointing. Use no more than two compass points for each direction.

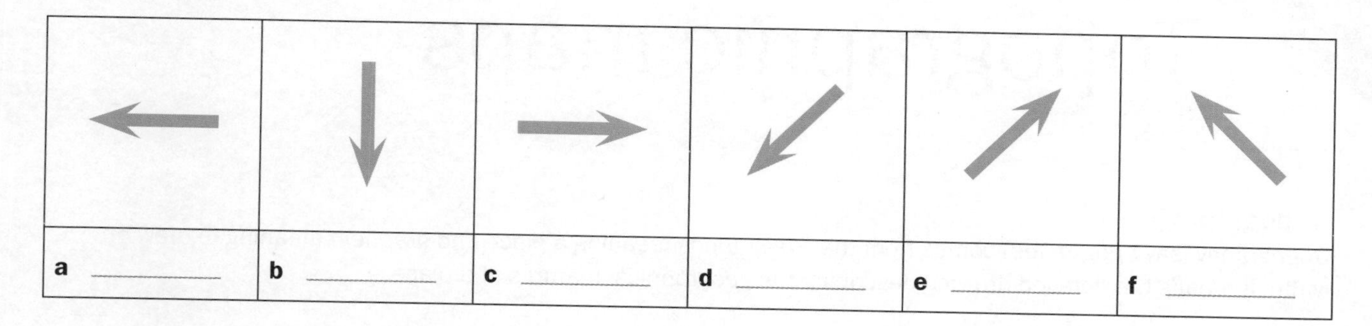

a ________	b ________	c ________	d ________	e ________	f ________

3 When Mt Tarawera erupted, the village of Te Wairoa was buried under hot, heavy ash and mud. The map shows what you see when you visit the Buried Village today. In each of the following six sentences, cross out the wrong blue answer to leave the correct one.

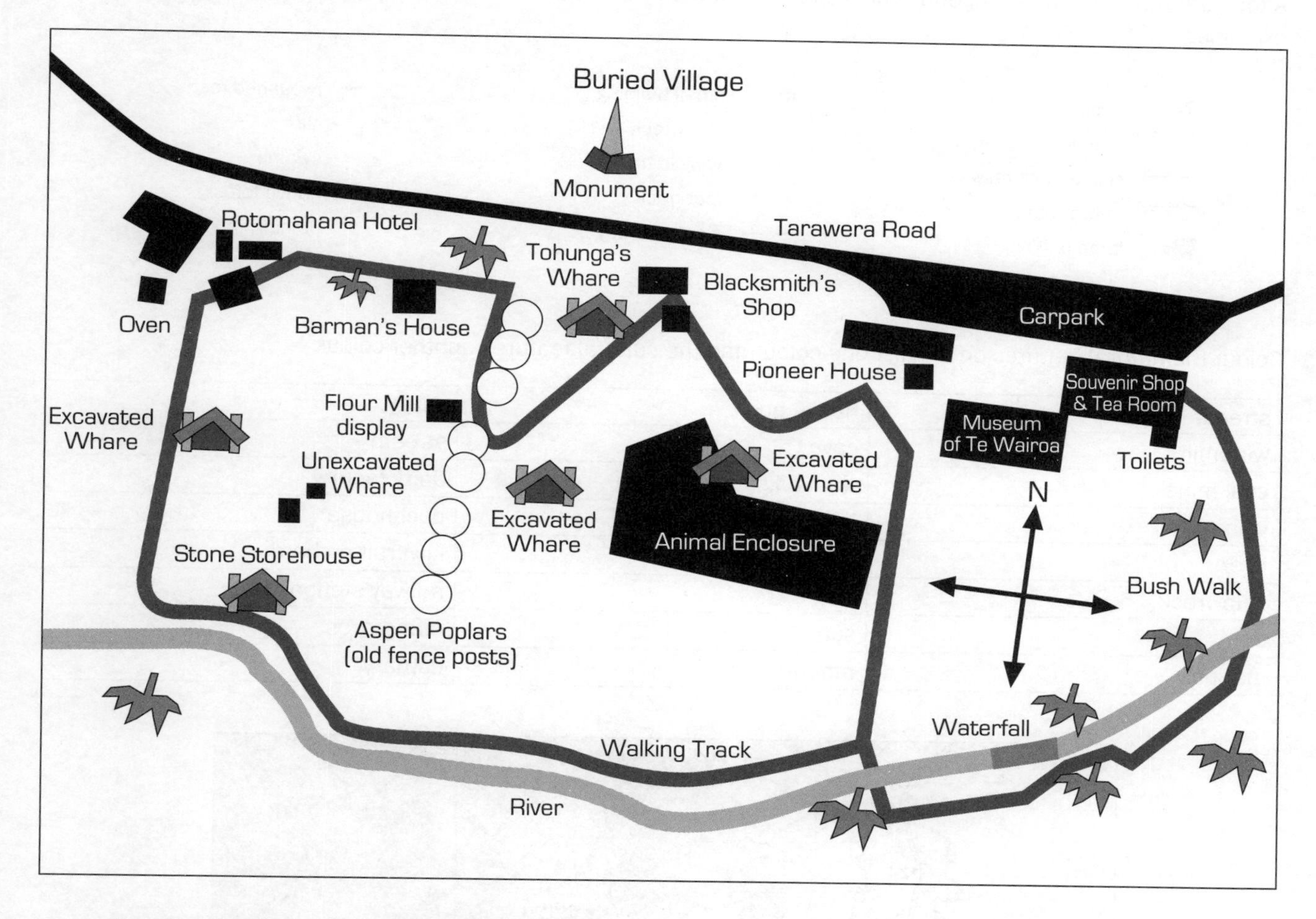

a The direction finder is **north/south** of the river.

b The Animal Enclosure is **east/west** of the direction finder.

c The Monument is **north/south** of the Excavated Whare.

d The Stone Storehouse is **east/west** of the Waterfall.

e The Blacksmith's Shop is **northeast/northwest** of the Flour Mill Display.

f The old fence posts are **southeast/southwest** of the Pioneer House.

ISBN: 9780170368131

Topographic maps

Topography

Topography (say *t'pog-ra-fee*) comes from the Greek *topos* meaning a place and *graphein* meaning to draw or write. It means the detailed drawing describing the geographical features of a place.

A topographical map is about the land and its features.
 Features can be cultural features, which are features made by people, such as schools.
 Features can be natural features, which are features made by nature, such as glaciers.

A topographical map has a legend, which shows what the symbols on the map stand for. Below are some examples.

- airport
- wind farm
- single-track railway
- railway station
- large building
- small building
- shipwreck
- vehicle track
- foot track
- state highway
- two-laned road
- cliff
- swamp
- native forest
- orchard or vineyard

1 Colour the natural features on this list one colour and the cultural features another colour.

streets	native bush	valley
windmill	cave	stream
coal mine	waste disposal system	dairy factory
volcano	lake	packhouse
river	cool storage	kiwifruit orchard
shipwreck	port	railway station
dam	sea	mangroves
marina	mountain	estuary

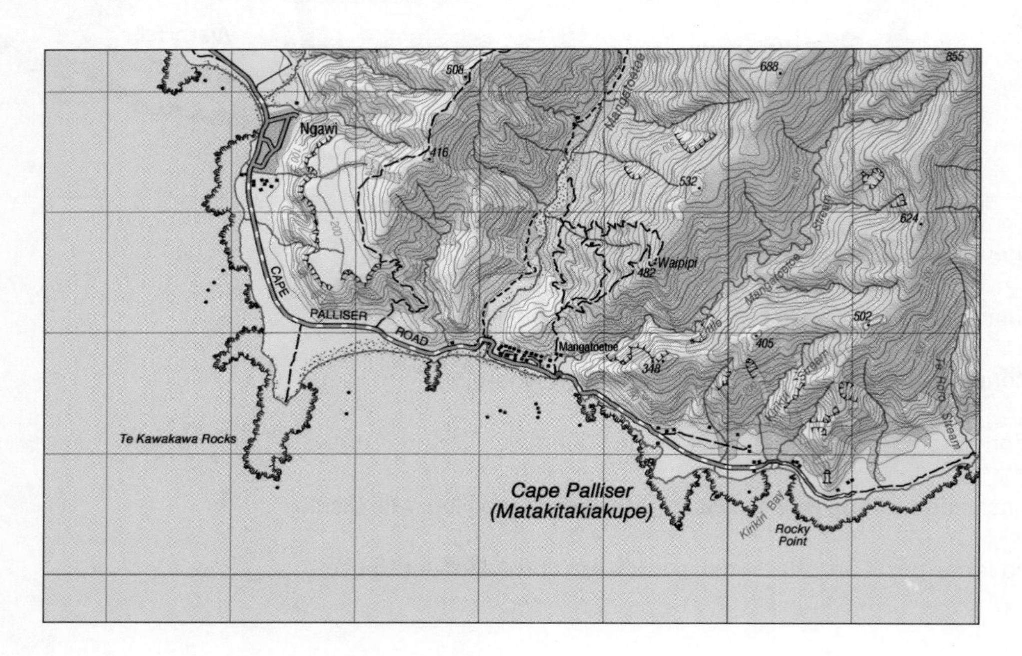

ISBN: 9780170368131

2 Look at the topographic map on page 22 and write answers to the following.

a What colour is used to show inland water? ______________________

b What do the numbers on the map stand for? ______________________

c Name a cultural feature at Ngawi and draw its symbol. ______________________

d Name a cultural feature around Rocky Point and draw its symbol. ______________________

e Where does the coastal vehicle track from the east become a road? ______________________

f What cultural feature would you see on your left when you went to Rocky Point from Ngawi?

____

g What does the colour green stand for? ______________________

h Is there evidence of a wind farm? ______________ An airport? ______________

3 Look at the topographic map on the inside of the front cover and give names for the following.

a The largest settlement. ______________________

b Two state highways. ______________________

c A major river. ______________________

d Ten cultural features. ______________________

____

e Five natural features. ______________________

____

f Where water from a power station discharges. ______________________

Grid references

To help you find things quickly, a map might have a grid.

A grid on a map is a set of parallel lines going up and down, and across.

Vertical grid lines go this way. Vertical grid lines are called eastings.

Horizontal grid lines go this way. Horizontal grid lines are called northings.

Together, vertical and horizontal lines make a grid pattern.

Grid lines are numbered, so you can look up the grid references for a place and use them to quickly find the place on a map.

Vertical easting lines are numbered from left to right — from west to east.

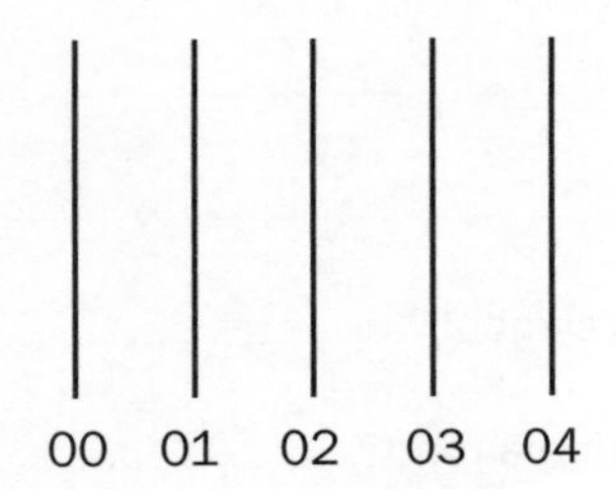

Horizontal northing lines are numbered from bottom to top — from south to north.

04
03
02
01
00

A grid reference usually has six figures.
The first three figures are for the easting.
The other three figures are for the northing.

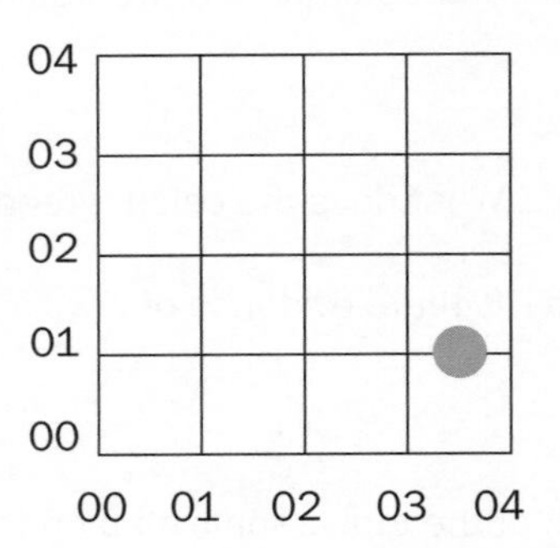

The orange dot is halfway between 03 and 04 easting. So its easting reference is 035, as the distance between 03 and 04 is measured in tenths.
The orange dot is right on the 01 northing so its reference is 010.
This means the full grid reference for the red dot is 035010.

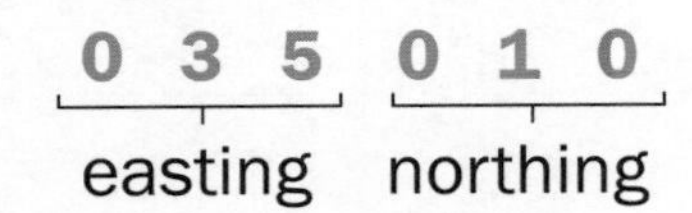

4 Beside each X on the map below, write the name of the feature the X refers to. Work out the answers by using the following grid references.

045140 = beacon
014115 = airstrip
020140 = marae
042110 = kiwifruit orchard
025124 = cemetery
028145 = historic Maori pa
044115 = cool-storage sheds
033128 = trig
015134 = dairy factory

Whenua

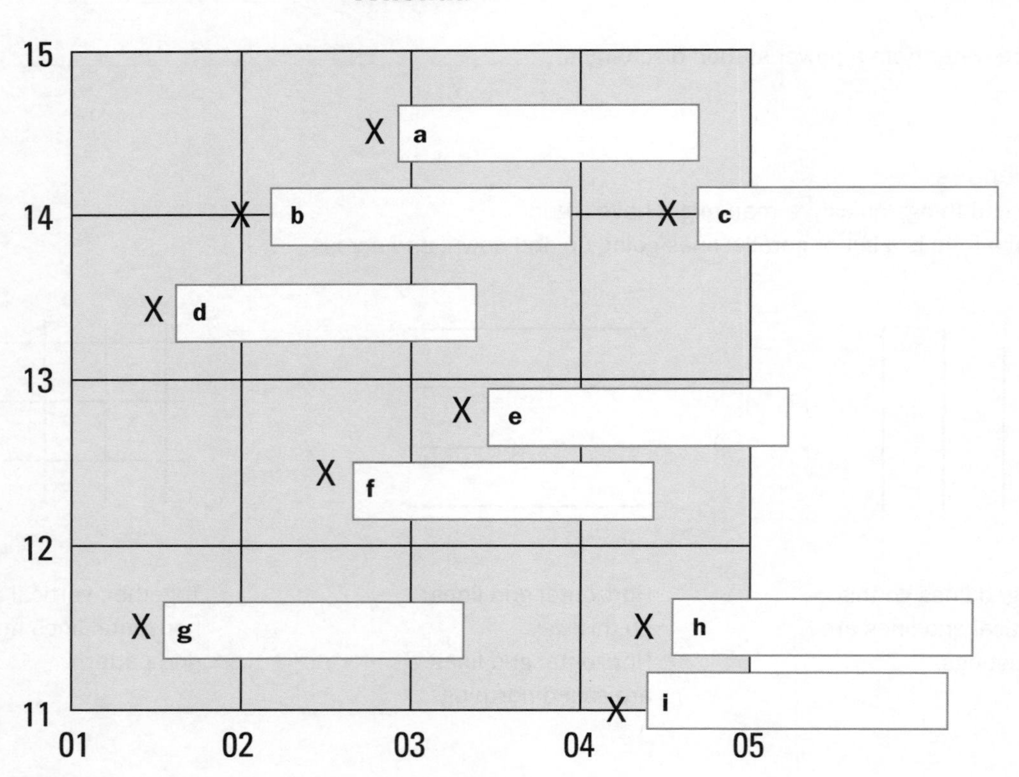

ISBN: 9780170368131

Map symbols

Using symbols on maps makes it easier and quicker to create maps and to read them. Draw and colour the topographic symbols in the boxes. Use the legend on the inside front cover.

a railway yard	**b** tunnel	**c** church

d historic Maori pa	**e** shelter belt	**f** large building

g small building	**h** power line on pylons	**i** two-lane bridge

j wharf, jetty	**k** exotic forest	**l** orchard

m native forest	**n** cemetery	**o** homestead

p residential area	**q** height of 100 m	**r** opencast mine

s swamp	**t** coastal rocks	**u** foot track

ISBN: 9780170368131

13 Latitude and longitude

Lines of latitude run parallel to each other around the earth horizontally.

North Pole 90° N
Equator 0°
Lines of latitude
South Pole 90° S

The equator divides the world into two hemispheres – the northern hemisphere and the southern hemisphere.

The equator is the 0 degree line of latitude.

Northern hemisphere
Equator 0°
Southern hemisphere

The other lines of latitude go north and south of the equator. The North Pole is 90 degrees north and the South Pole is 90 degrees south.

North Pole

Lines of longitude run from the North Pole to the South Pole.

Lines of longitude

This is the Earth flattened to show the divisions between longitude lines. The 0 degree line and the 180 degree line divide the world into two hemispheres – the eastern hemisphere and the western hemisphere.

180° 150° 150° 120° 120° 90° 90° 60° 60° 30° 0° 30°
West East

A longitude line is always east or west, except for the 0 degree line and the 180 degree line because there is only one of each.

Greenwich in England
Western hemisphere
Eastern hemisphere
0° longitude

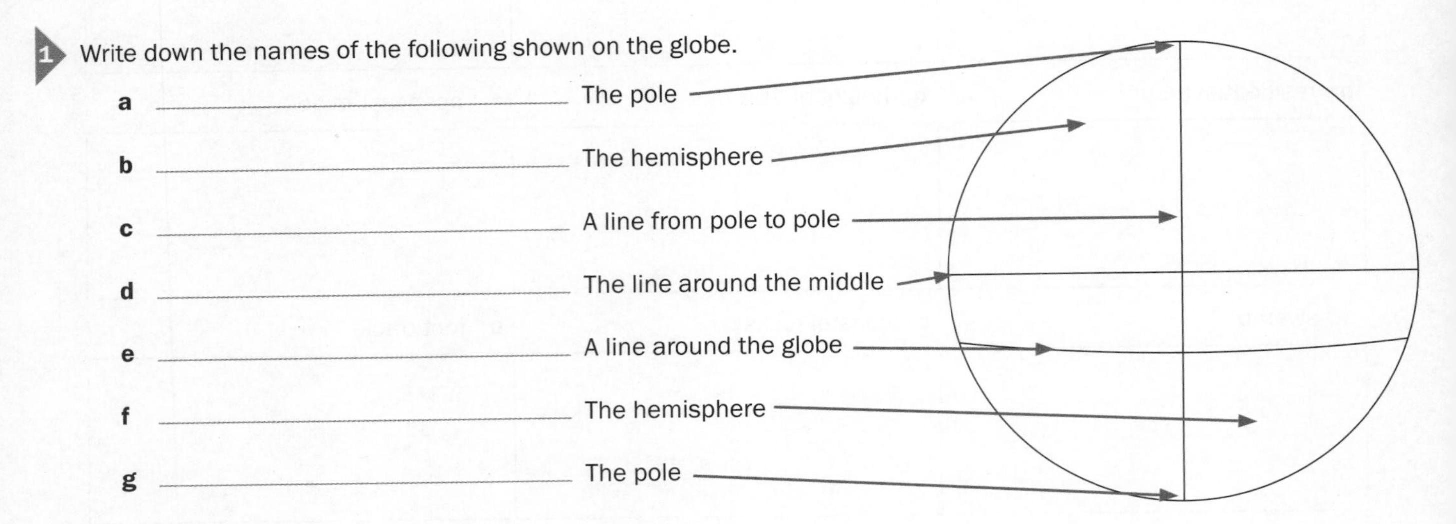

ISBN: 9780170368131

2 Look at the map and fill in the following about it.

a The numbers down the left side refer to lines of ____________

b If the next line was drawn after 10, it would be the ____________

c The numbers across the top refer to lines of ____________

d The latitude line through New Zealand is ________ degrees.

e The longitude line through the Cook Islands is ________ degrees.

f The latitude line through Tonga is ________ degrees.

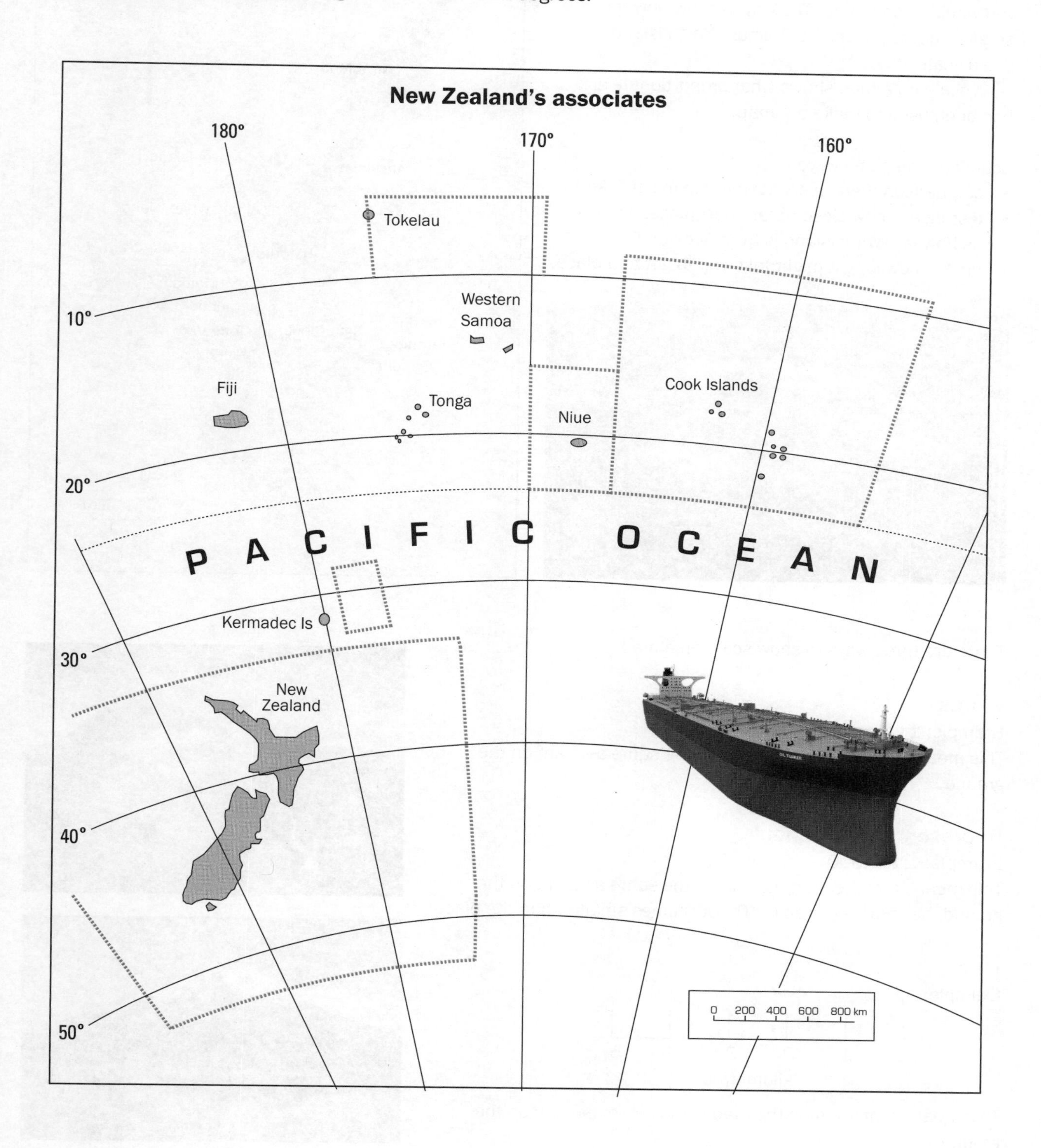

ISBN: 9780170368131

14

Scale and distance

Scale on a map means the proportionate (relative) size of something.

If you wanted to explore the major volcanoes in New Zealand, which are all in the North Island, you would need a map. The way to show how much smaller the map is than the actual North Island is to use a scale.

A scale on a map shows what proportionate size has been used to make the map.

Scale on a map shows you

- comparison (here every 12 mm means 100 km)
- distance – how close or far apart places are such as how far White Island is from Auckland
- time – how long it might take you to get to a place.

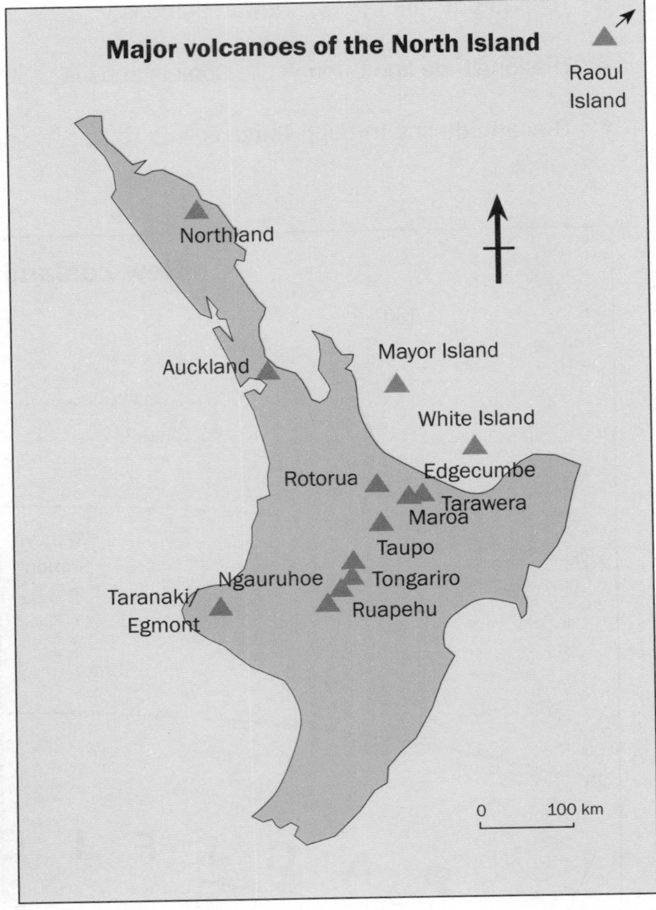

There are three ways to show scale on a map.

Words

Example: 1 cm = 1 km

This means that 1 cm on the map is the same as 1 km on the ground.

Representative fraction

Example: 1:100,000

This means that 1 cm on the map is the same as 1 km on the ground, because the map is 100,000 times smaller than the place.

Line

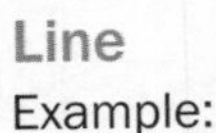

Example:

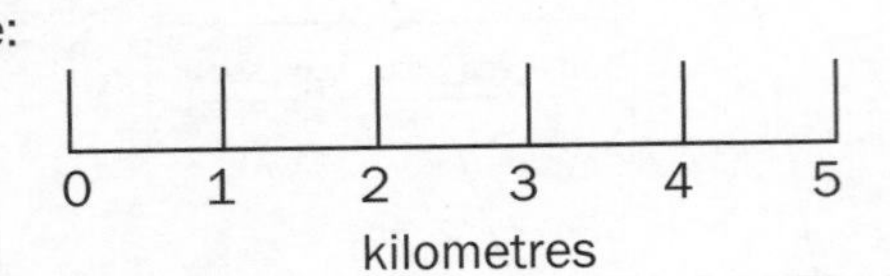

This means that 1 cm on the map is the same as 1 km on the ground.

 ISBN: 9780170368131

1 Look at the map on page 28.

a Write down the way the scale is shown.

____

b Write down another two ways the scale could have been shown.

____

____

2 Look at the map of Volcano Island below and write down the scale in two different ways.

____

____

3 Use the scale to work out these distances on Volcano Island in a straight line.

a A to B ________ **b** B to C ________ **c** C to D ________ **d** D to E ________

e E to F ________ **f** F to A ________ **g** A to D ________ **h** C to F ________

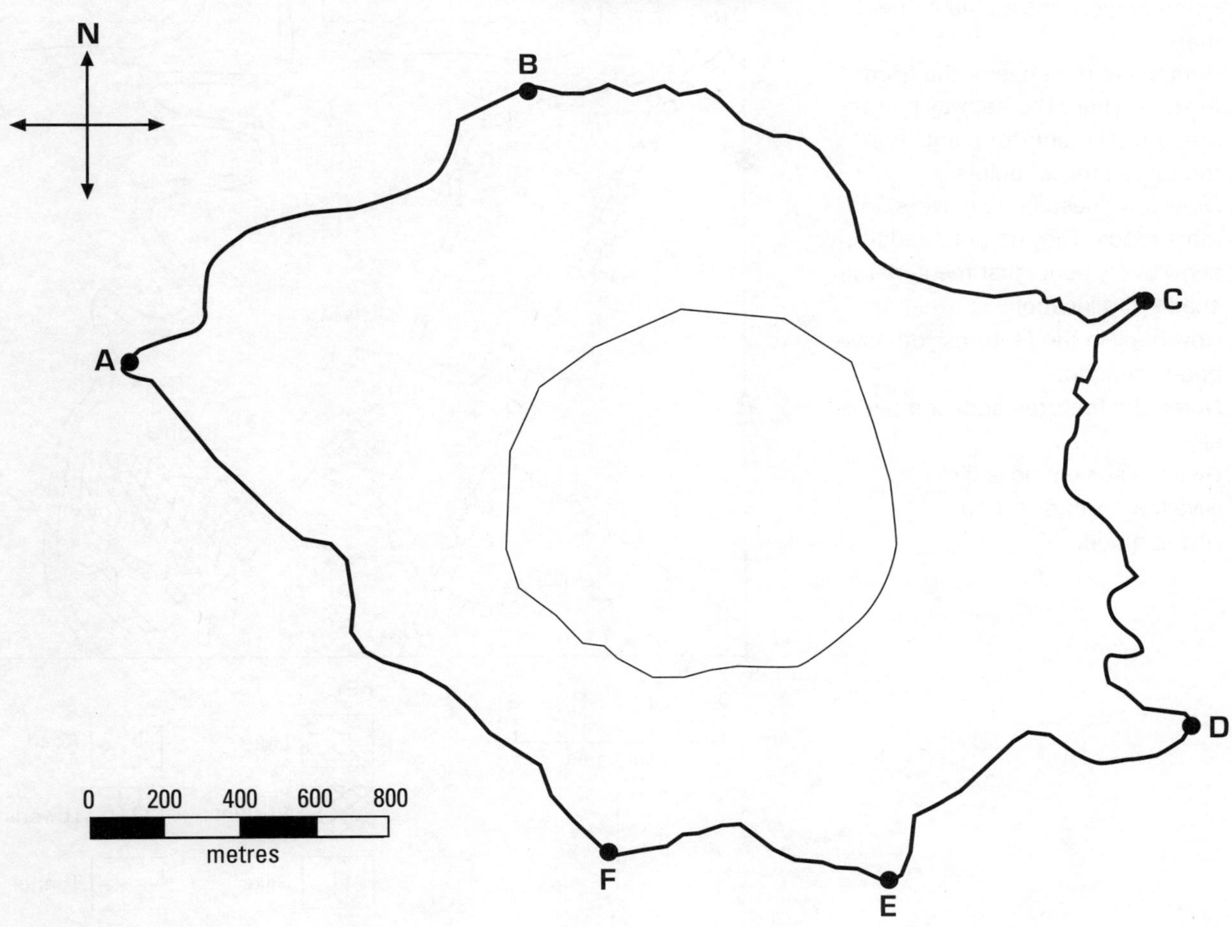

ISBN: 9780170368131

15

Precis maps

Precis (say *pray-see*) = summary = picking out the main points.

Precis map = summary of another map = picking out the main points of the map.

Although there are wonderful places all around the Turangi area such as hot springs, a precis map does not show these individual places because it is a summary of the main points.

Main rules about drawing a precis map

1. Work out exactly what you have to show and what you have to leave out.
2. Make it simple.
3. Make it neat.
4. Make it big and clear.

How to draw a precis map

1. Use a ruler to draw a frame in the same proportions as the actual map.
2. Put dots at the edge of the frame to show where the halfway points are, then the quarter points and the three-quarter points.
3. Draw any coastline, big rivers and main roads. They do not need to show every bend, just the general shape and location.
4. Now draw in the features you have been asked to.
5. Name the features and/or make a key.
6. Give it a scale if possible.
7. Give it a north direction.
8. Give it a title.

Precis map of an area of Turangi

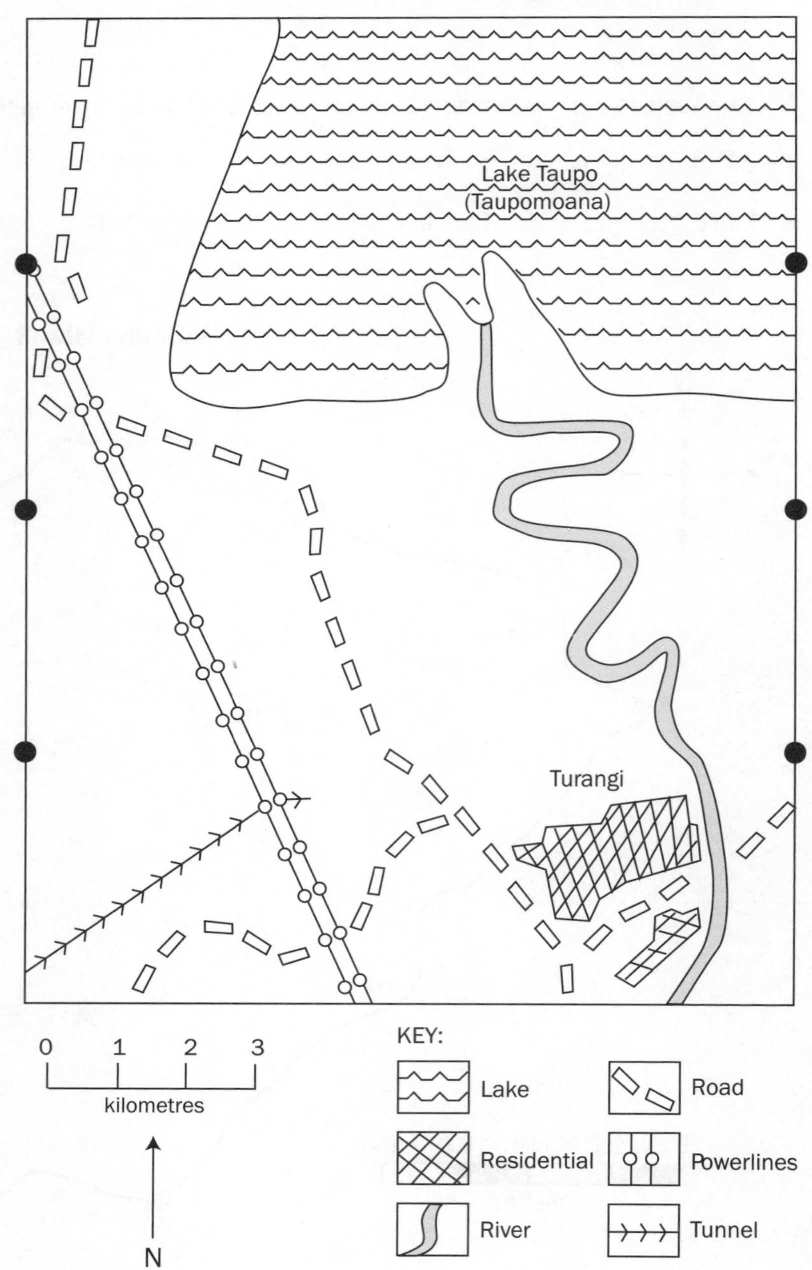

 ISBN: 9780170368131

1. Put your own colour on the precis map of Turangi.

2. In the map below, Li has located and placed all the features he was asked to on a precis map, but made five mistakes with his map. List the mistakes.

   a ____________________

   b ____________________

   c ____________________

   d ____________________

   e ____________________

TOWNSHIP

NATIVE FOREST

DAIRY FARMING

RAILWAY

KIWIFRUIT ORCHARDS

HIGHWAY

RIVER

3. In the blank box, draw a precis map of the following photograph of Mount Maunganui taken from Mauao, the mountain.

16

Contour lines

Relief means how the land is shaped, whether it is flat, hilly or mountainous, whether there are plains or valleys and so on, and how high those shapes are.

One way to show relief on a map is to draw contour lines. Contour lines join places of equal height.

This is how the hill looked from sea level.

This is how the scene looked from above.

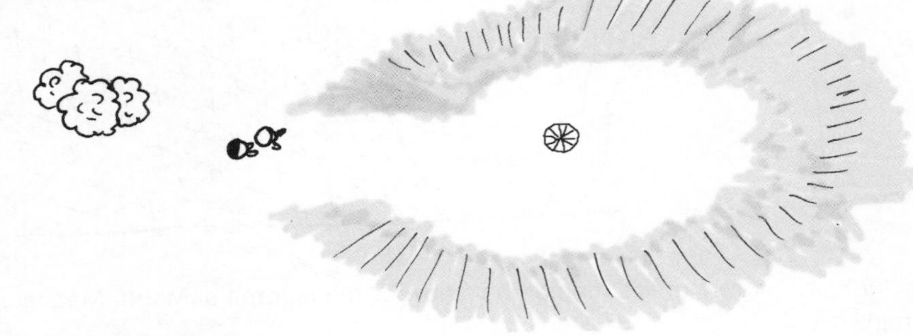

This is how the hill looked on a topographical map.

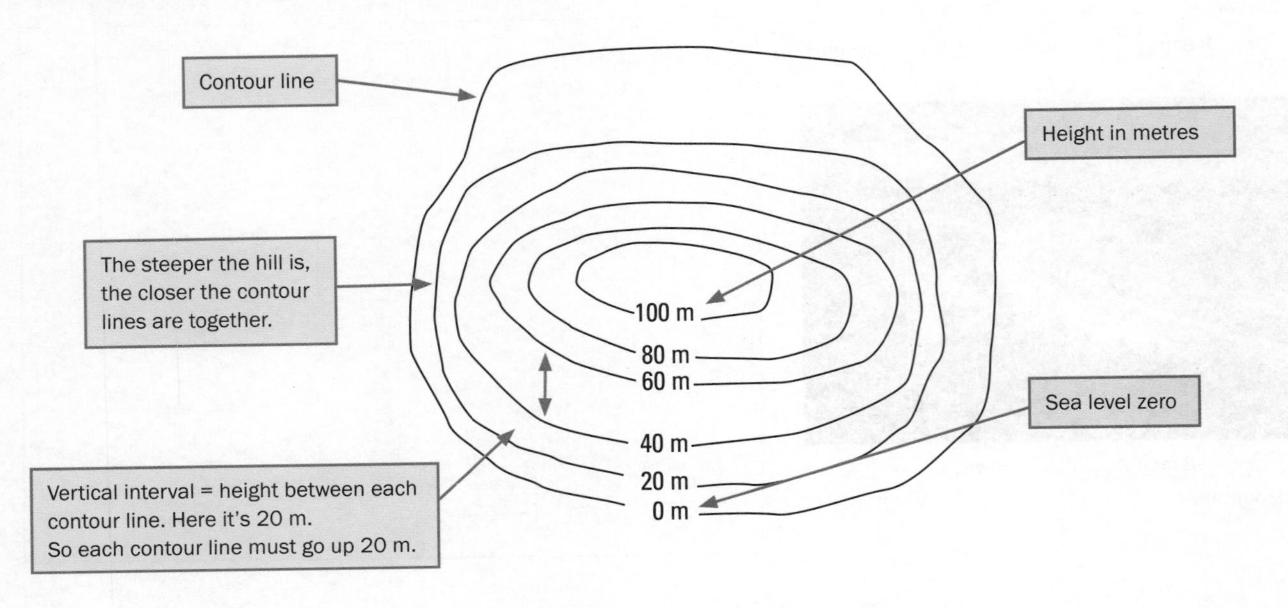

ISBN: 9780170368131

Contour lines of common features

Hills and mountains have contour lines that get higher towards the middle.		
A **steep slope** has contour lines close together because the height goes up quickly over a short area.		
A **valley** has a V-shaped pattern on its contour line. The Vs point uphill.		
A **ridge** has a V-shaped pattern on its contour lines. The Vs point downhill.		
A **depression** (hole) has contour lines with heights getting lower instead of higher. Contour lines have small marks on them pointing towards the middle.		

1 Give the heights of the points marked.

a ________

b ________

c ________

d ________

e ________

f ________

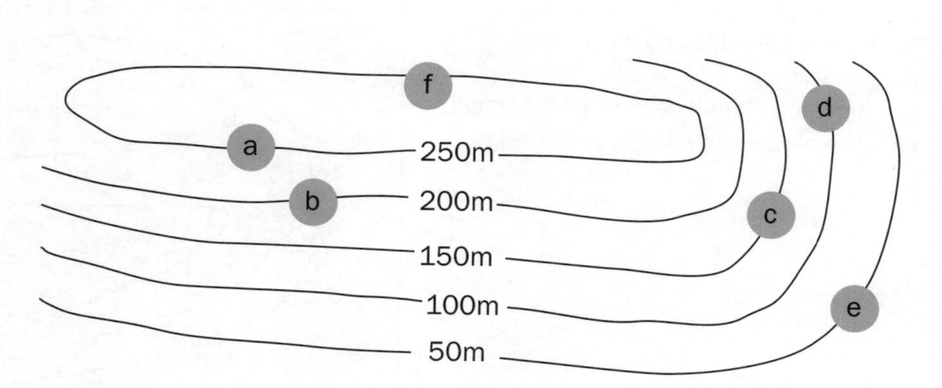

2 Write down the name of the feature shown by the contour lines.

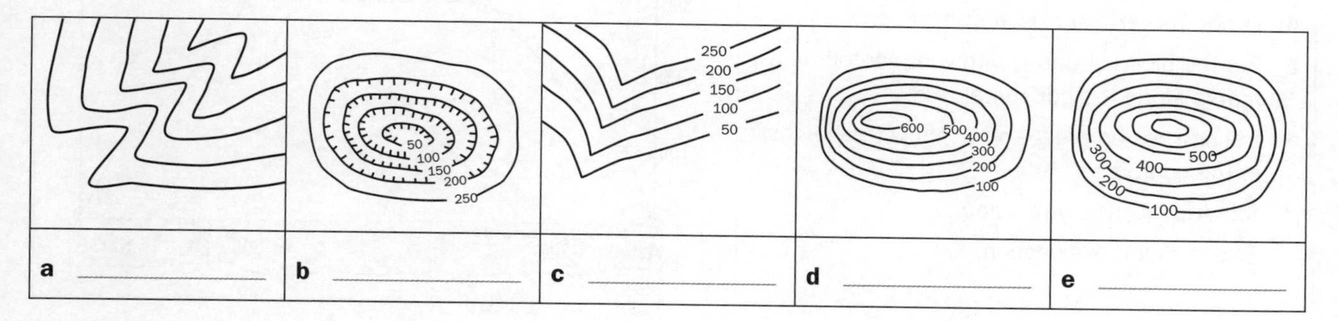

ISBN: 9780170368131

17

Cross-sections

A cross-section gives a picture of the relief of the land along a particular line as if a giant knife has cut through the land, scraped away one half and left you looking at the side of the half that is left.

You can draw a cross-section by using the contour lines on a topographic map.

Aim: To draw a cross-section of the island along the A–B line.	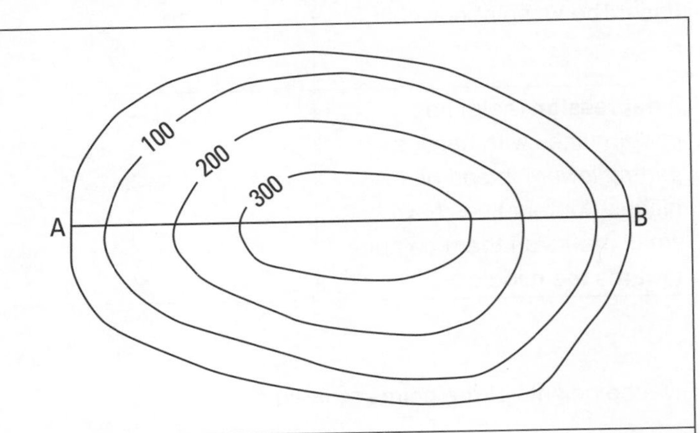
1 Put the straight edge of a piece of paper along the line. **2** Each time a contour line crosses the paper, mark the paper with a line and the height.	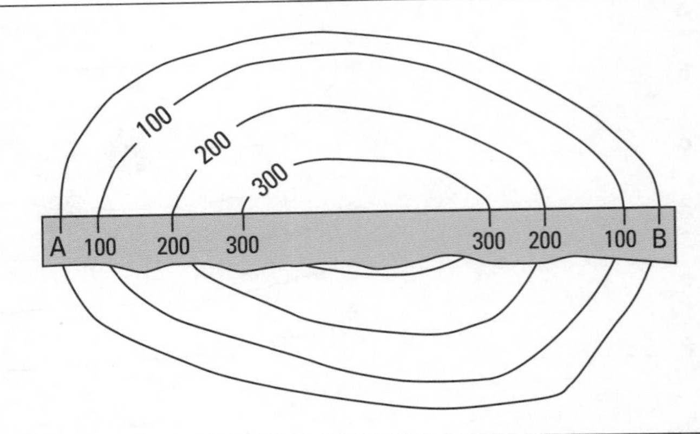
3 Draw a vertical and horizontal axis for a graph. **4** Label the vertical axis in metres. **5** Put the piece of paper with your contour line marks along the horizontal axis. **6** Put crosses on the graph to show the heights of the contours. **7** Join the crosses with a line. This is your cross-section.	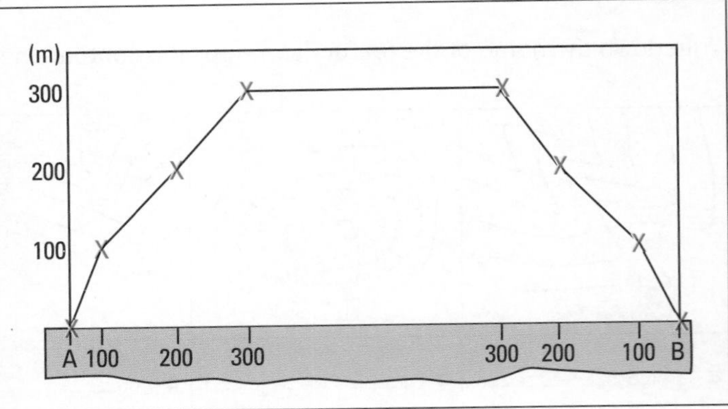

 ISBN: 9780170368131

1 Label the following cross-sections with 'mountain', 'river', 'plain' or 'plateau'. Add colour.

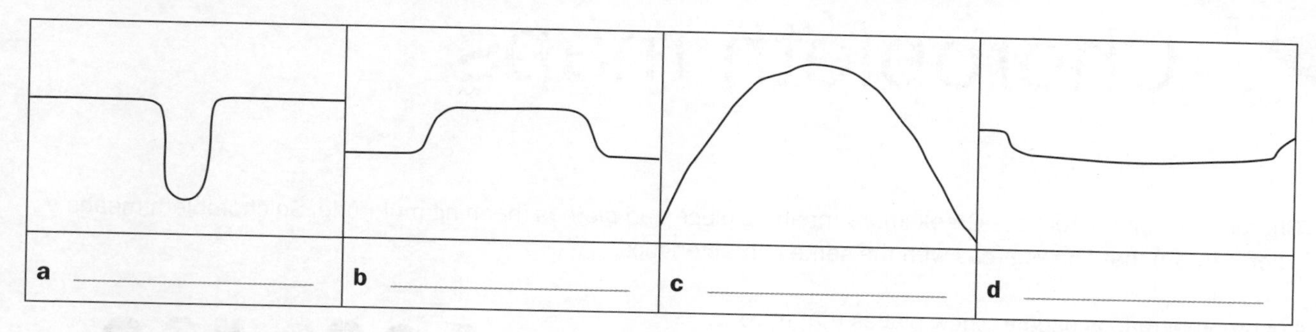

2 a Mark in the heights along the piece of paper that are needed to draw a cross-section.

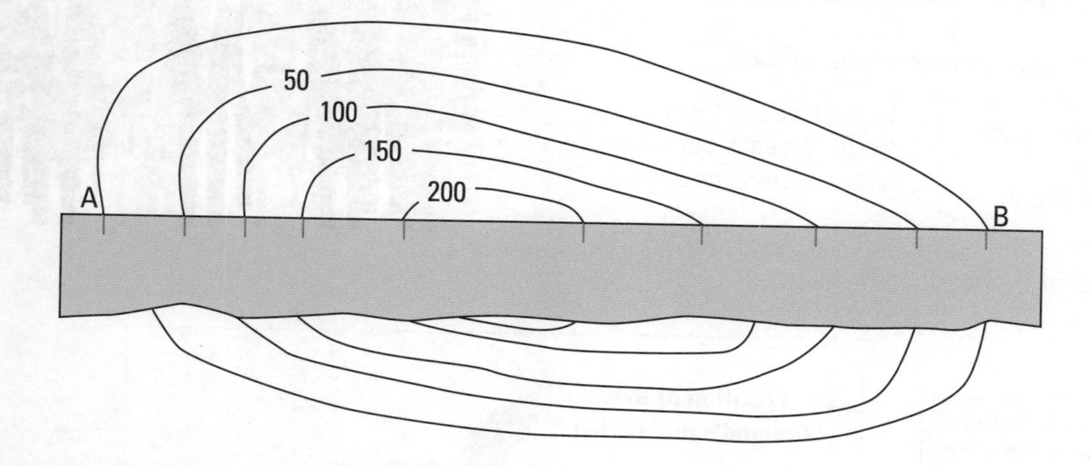

b Label the vertical axis on the frame below and use the marks on the strip of paper to draw your cross-section.

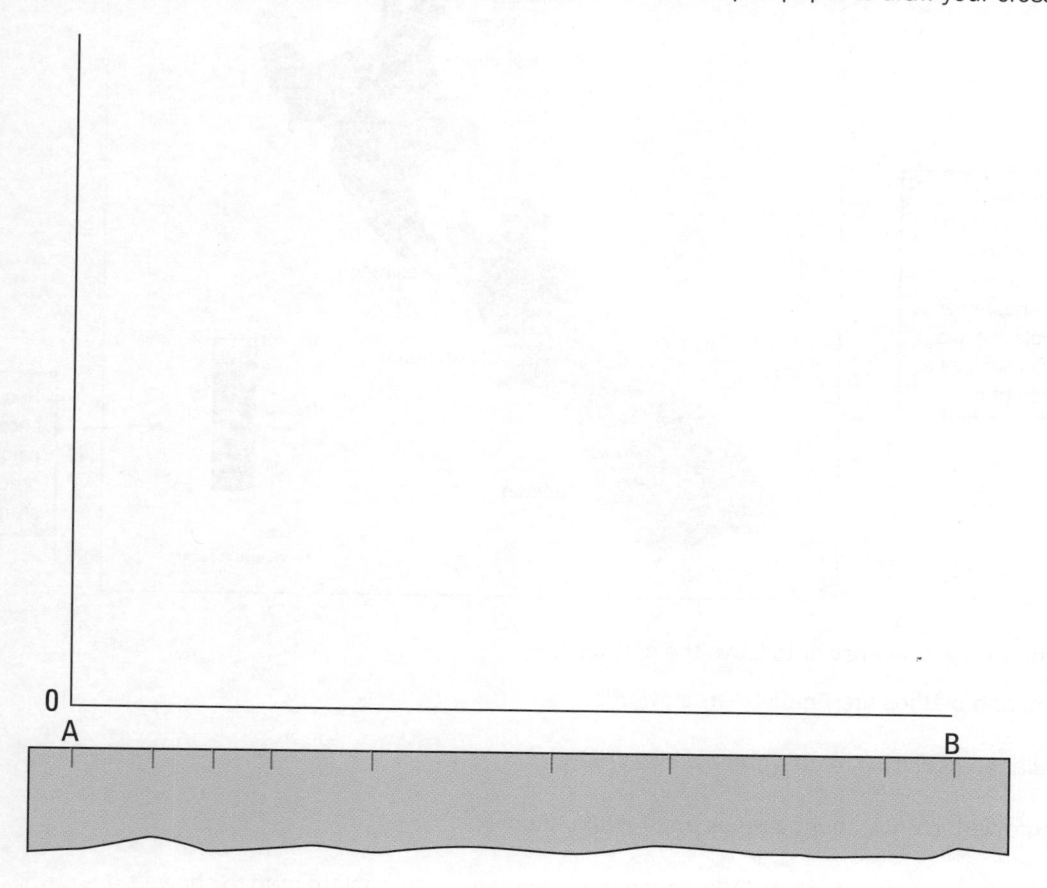

ISBN: 9780170368131

18

Choropleth maps

'Choropleth' comes from the Greek *khora* meaning place and *plethos* meaning multitude. So choropleth means a graphic (map) that shows areas with the same characteristics.

A choropleth map is used to show places that have
- the same thing, such as people living there (population)
- different amounts of the same thing, such as China having a larger population than New Zealand.

Common things a choropleth map shows:
- Population
- Sunshine hours
- Rainfall
- Temperatures

} Each place has these things but each place has different amounts of them.

Example of a choropleth map

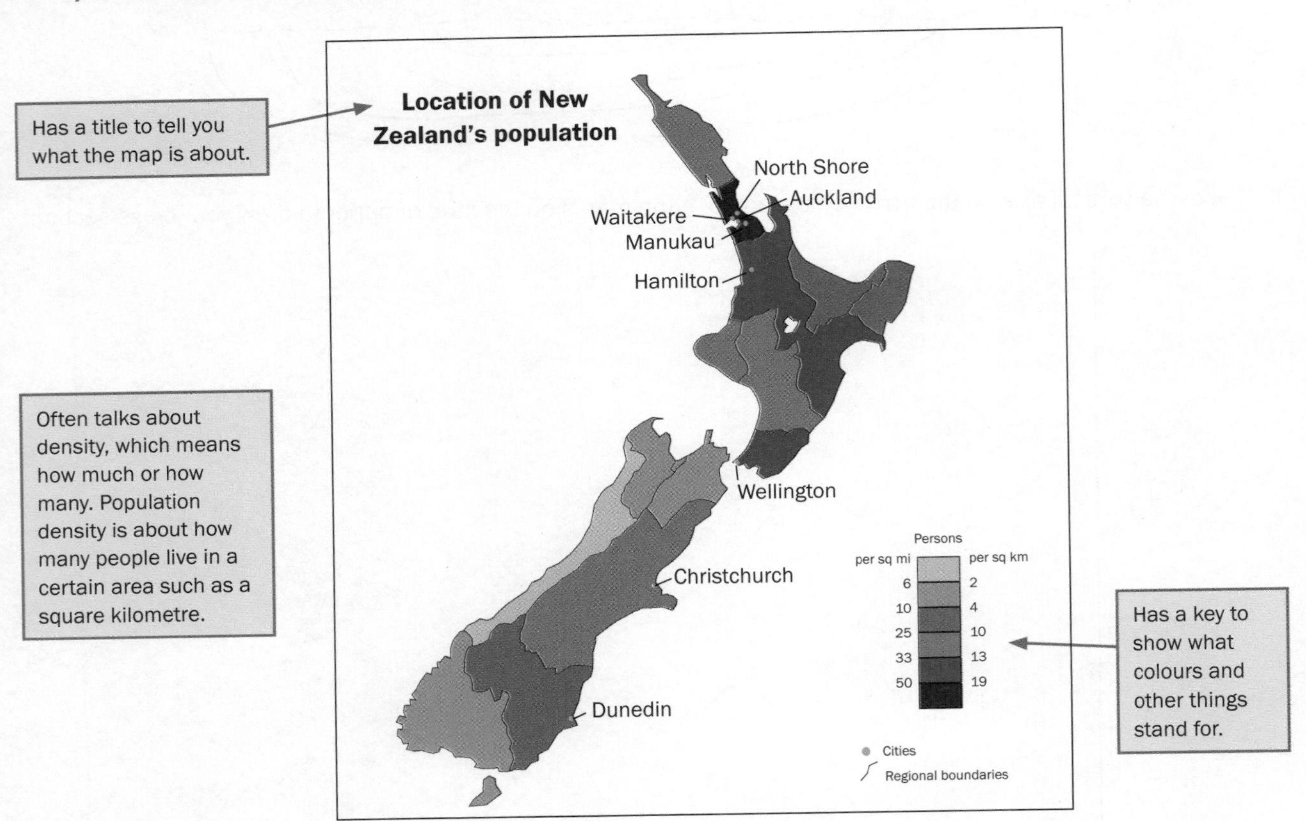

1 Cross out the wrong blue answer to leave the correct one.

a *Khora* and *plethos* are **English/Greek** words.

b Population density is to do with **how many/why**.

c A choropleth map is about **causes and results/places**.

d Rainfall is **a common feature/an uncommon feature** for a choropleth map to show.

ISBN: 9780170368131

Complete the choropleth map below by adding colour to the map and a key.

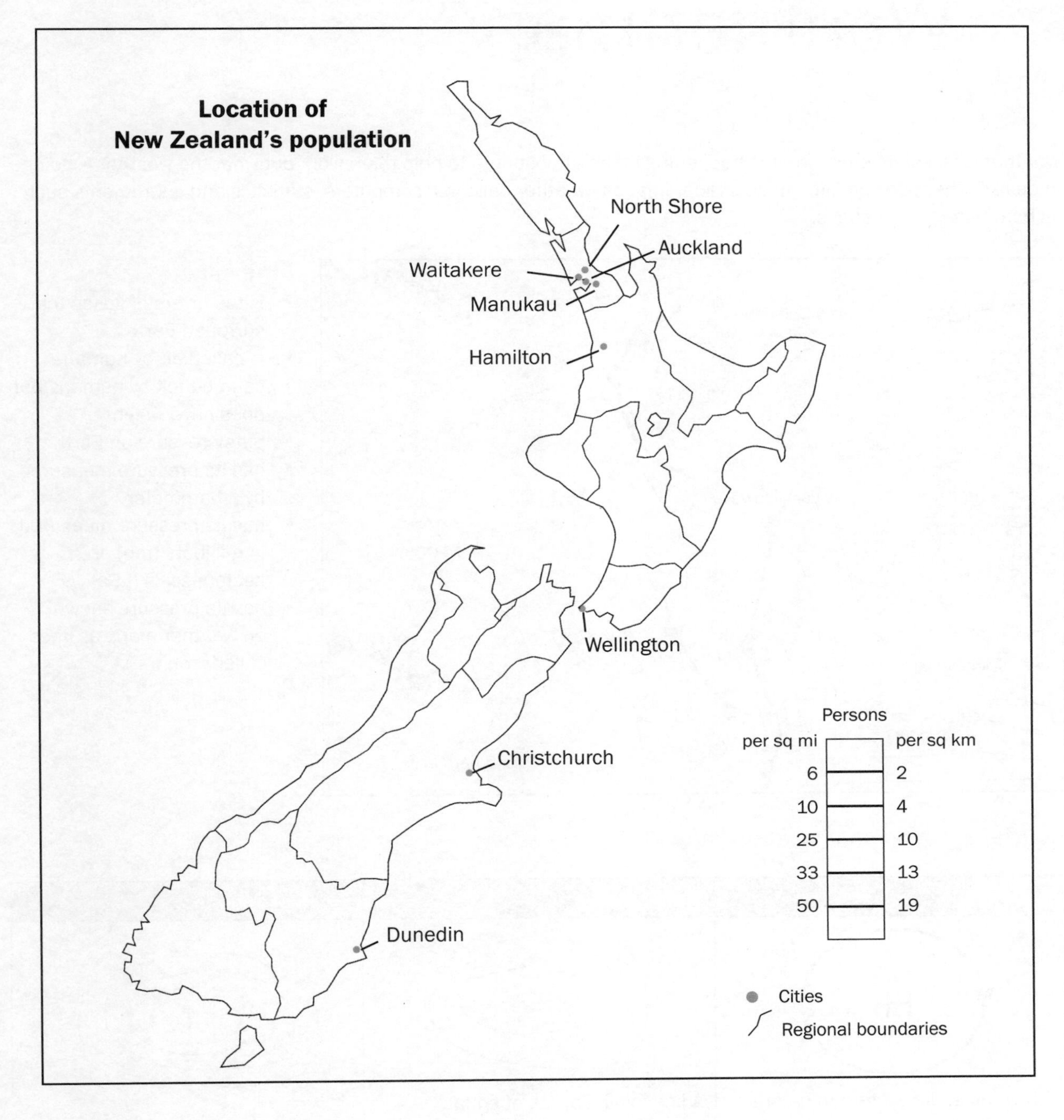

3 Put a tick or cross beside each of the following to show which are true and which are incorrect.

- ☐ **a** Population is spread unevenly over New Zealand.
- ☐ **b** More people live in the South Island than in the North Island.
- ☐ **c** Population density numbers are different if they are given in square miles rather than square kilometres.
- ☐ **d** The population density is greater in Auckland than in the East Coast of the North Island.
- ☐ **e** Most of the South Island has a population density of fewer than two persons per square kilometre.

ISBN: 9780170368131

19

Weather maps

Weather forecasters study what is happening in the atmosphere to help them work out what the weather is going to be like. They use satellite images, radar images, weather balloons, computers, statistics and instruments such as barometers and gauges.

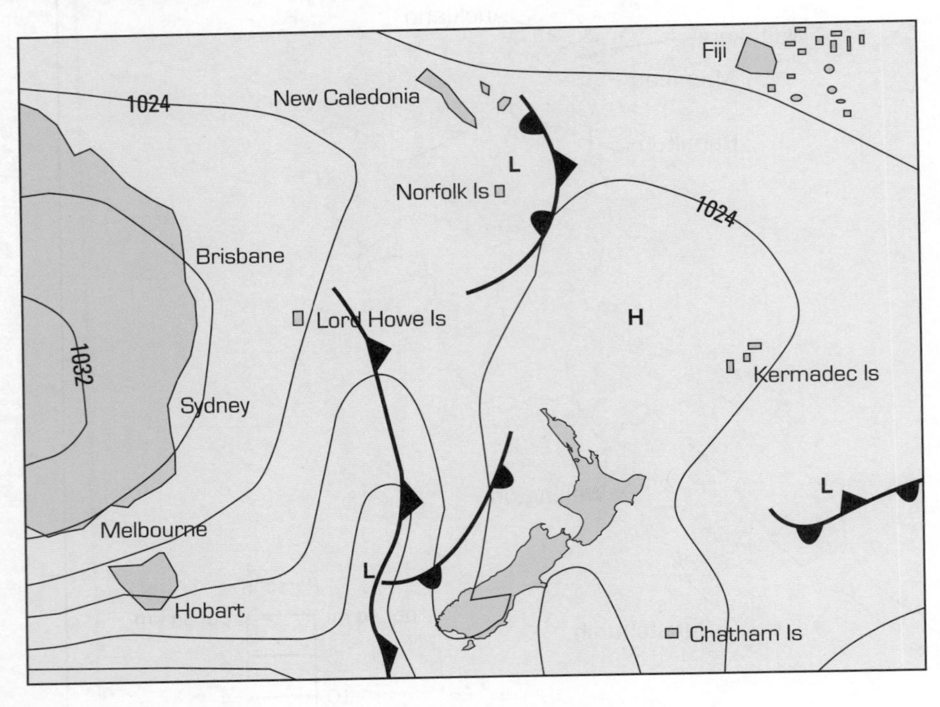

Atmosphere

- is the layers of gases that surround Earth
- is called air by humans
- can't be felt by humans but does have weight
- puts pressure on Earth
- has its pressure measured by a barometer
- has its pressure measured in millibars (mb) or hectopascals (hPa)
- has its pressure shown on weather maps as lines called isobars.

Special symbols on a weather map

Anticyclone	Isobar	Tropical cyclone
H A high-pressure system that usually brings fine and calm weather; wind goes anticlockwise.	A line joining areas of equal pressure; isobars close together means wind.	T Very low-pressure zone; winds go clockwise; brings lots of rain.
Depression	**Pressure readings**	
L A low-pressure system that usually brings wet and windy weather; wind goes clockwise.	1000, 990, 980, L 1020, 1030, H Low pressure means bad weather; high pressure means good weather.	

ISBN: 9780170368131

Front (marks the boundary between warm air and cold air)		Winds
Cold front: often brings rain.	**Warm front**: often brings drizzle.	Southwesterly Northerly **Winds** are named after the direction from which they blow. For example, a wind blowing from the southwest is called a southwest wind or a southwesterly, while a wind blowing from the north is called a north wind or a northerly.
Occluded front: cold front catches up with warm front; often brings long, rainy sessions.	**Stationary front**: no movement of air; brings long continuous rainy sessions.	Fast wind Slow wind Fast winds are shown by isobars close together; slow winds are shown by isobars far apart.

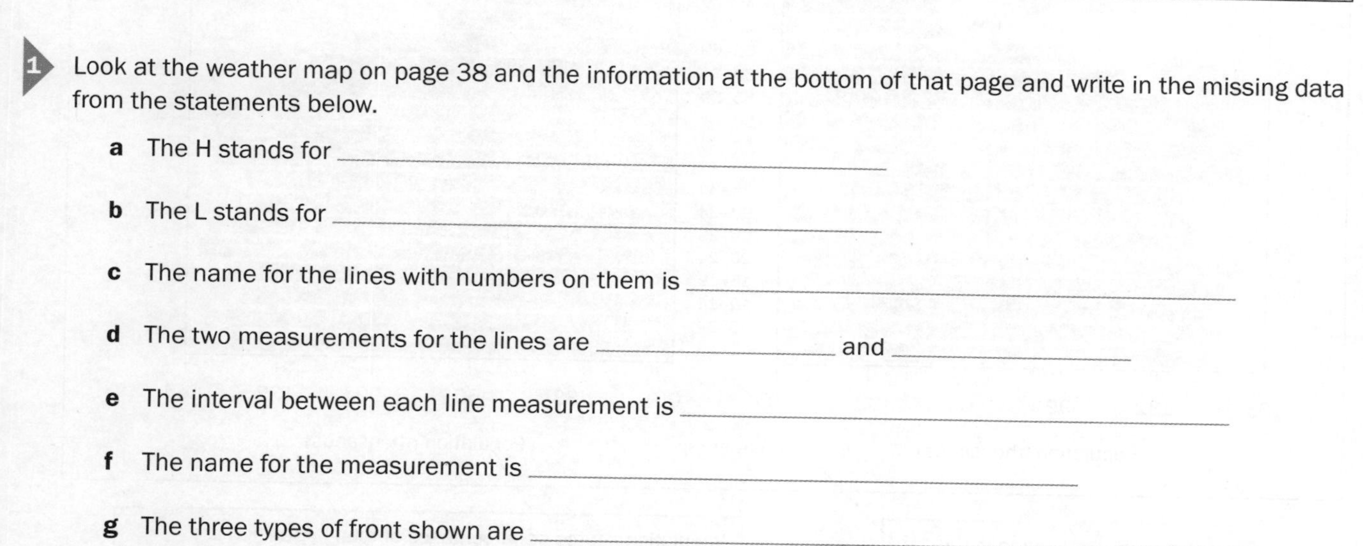

1 Look at the weather map on page 38 and the information at the bottom of that page and write in the missing data from the statements below.

a The H stands for ______________________

b The L stands for ______________________

c The name for the lines with numbers on them is ______________________

d The two measurements for the lines are ______________ and ______________

e The interval between each line measurement is ______________________

f The name for the measurement is ______________________

g The three types of front shown are ______________________

____

2 Cross out the wrong option in the following about the weather map.

a The weather in New Zealand is expected to be fine/rainy the next day.

b There should be a slow breeze/fast wind over New Zealand the next day.

c The map shows weather mainly for New Zealand/Australia.

d The isobars over New Zealand are close together/far apart.

ISBN: 9780170368131

20 Population pyramids

Population = people (it comes from the Latin word *populus*, which means people)

Pyramid = a shape with sloping faces that meet at the top.

Example of a population pyramid

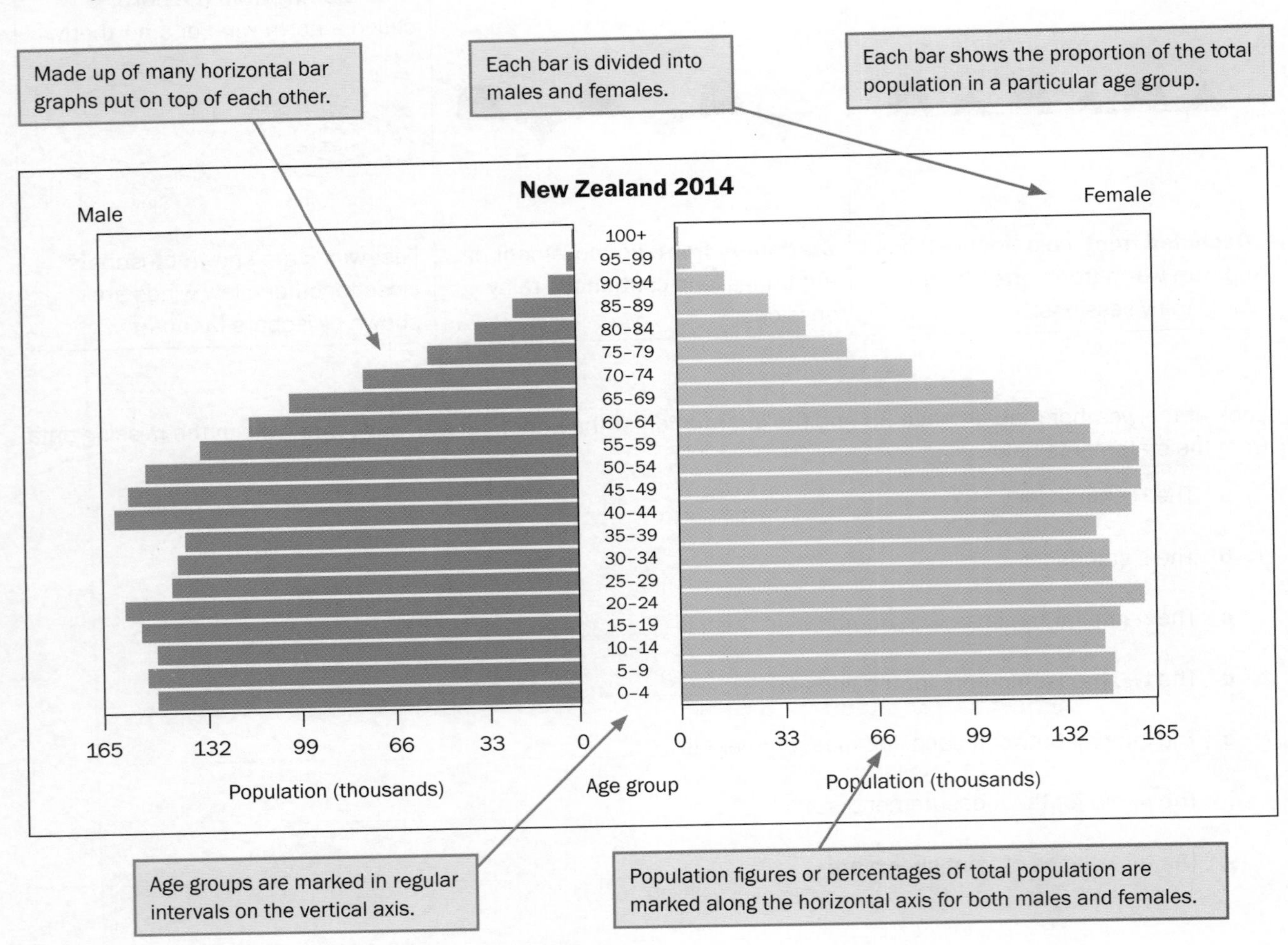

1 Look at the population pyramid of New Zealand and fill in the gaps in the following.

a The ages are marked on the ____________ axis and the population on the ____________ axis.

b The ages and population are marked at ______________________ intervals.

c Each bar is divided into ______________________ on the left and ______________________ on the right.

 ISBN: 9780170368131

2 Copy the population pyramid into the blank box. Use a ruler. Use your own colours.

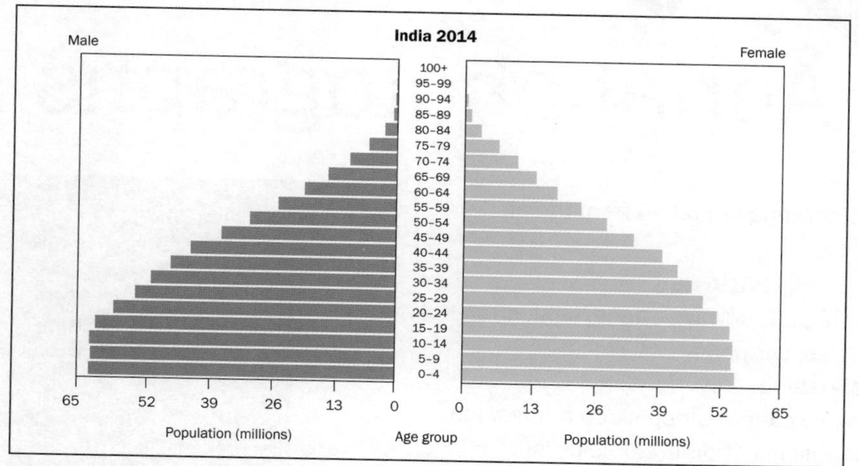

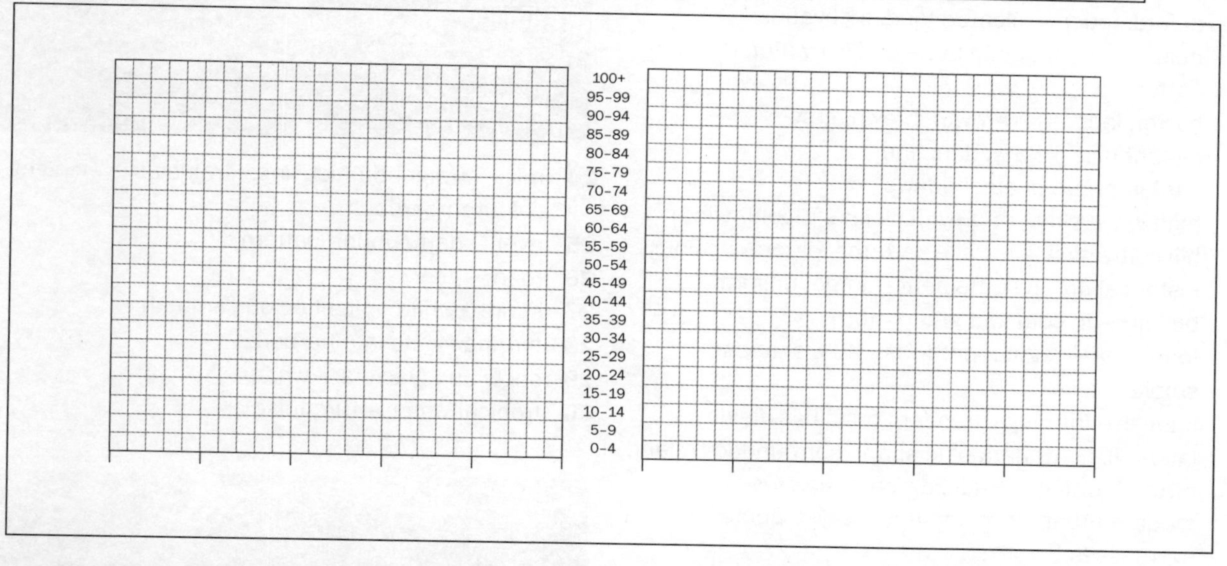

3 Look at the population pyramid for Bahrain and write down the following data about it. Round figures off.

a The number of females aged 20–24 is ____________________

b The number of males aged 20–24 is ____________________

c The total number of children aged 0–4 is ____________________

d The number of males aged 30–34 is ____________________

e The number of males aged 10–14 is ____________________

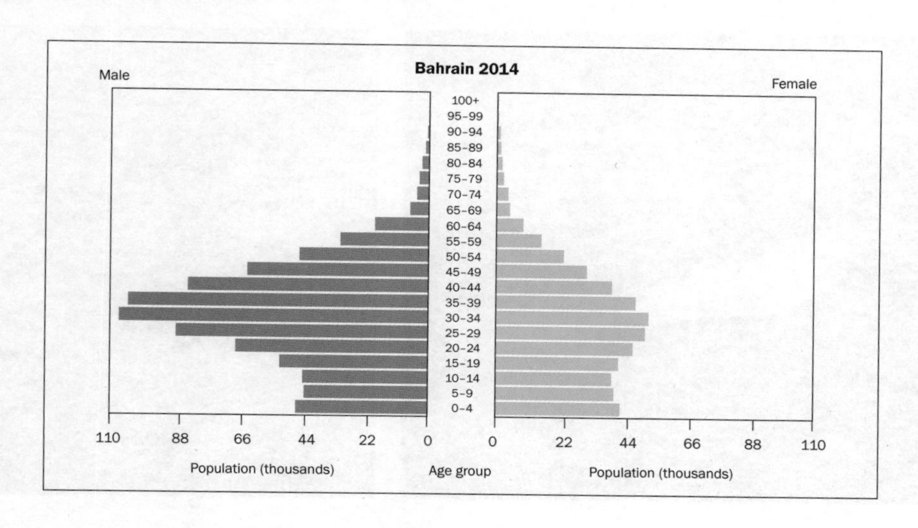

ISBN: 9780170368131

21

Aerial photographs

Aerial photographs are taken from the air.

A vertical angle = a bird's-eye view, directly overhead, often harder to pick out what's what in the photograph.

An oblique angle = taken from an angle, sloping rather than direct, easier to pick out what's what in the photograph.

Natural features often = curving, uneven, random patterns.

Cultural features often = straight, parallel, geometric, regular patterns.

Examples of features

1 airports and runways – long, light, straight lines, buildings, car parks
2 cliffs – steep, grey
3 crater/volcano – steep-sided depression
4 crops – fields of different colours
5 deforestation – cleared land, cultivation
6 delta – river meeting large body of water, fan-like pattern
7 barren land – no growth, brown, grey
8 distant hills – hazy, dark line
9 flat land – no deep shadows
10 highway – white or grey thick line, regular width
11 hill cultivation – terrace patterns, different colours
12 historical urban settlement – large, uniform buildings in compact grid pattern
13 industrial – regular patterns, white shapes, smokestacks
14 intensive farming – random buildings, fields
15 lake – light- or dark-blue shape surrounded by land
16 mixed farming – fields, different colours
17 modern urban development – skyscrapers, overpasses
18 mountains – steep, bold lines, shadows, valleys
19 native vegetation – different greens, dense, dark patches
20 ocean – dark blue
21 planned residential – houses uniform in shape and colour, curved streets, regular patterns
22 pollution – discolouration due to oil/chemical leak/discharge
23 quarry/mine – vegetation removed, light/white, terrace patterns, blue tailing ponds
24 rail – straight darkish lines, trains as even, light-coloured boxes
25 river – dark, ribbon pattern
26 road – grey
27 sand – white or light colour, smooth
28 marina – juts into harbour
29 snow and glaciers – white
30 tropical water – turquoise

 ISBN: 9780170368131

1 Look at the 30 examples of features and 11 aerial photographs. Find an example in the aerials of each feature.

ISBN: 9780170368131

22

Precis sketches from aerial photographs

Background

Foreground

A precis sketch from a vertical photograph is sketched inside an ordinary square or rectangle because the photograph is taken straight down and so the foreground is the same measurement as the background.

Because an oblique photograph is taken at an angle, the foreground is shorter than the background. This means that a precis sketch from an oblique photograph has to be sketched inside a trapezium frame. The trapezium shape stretches out the background and shortens the foreground.

Background

Foreground

How to draw a precis sketch from an aerial photograph (see inside back cover)

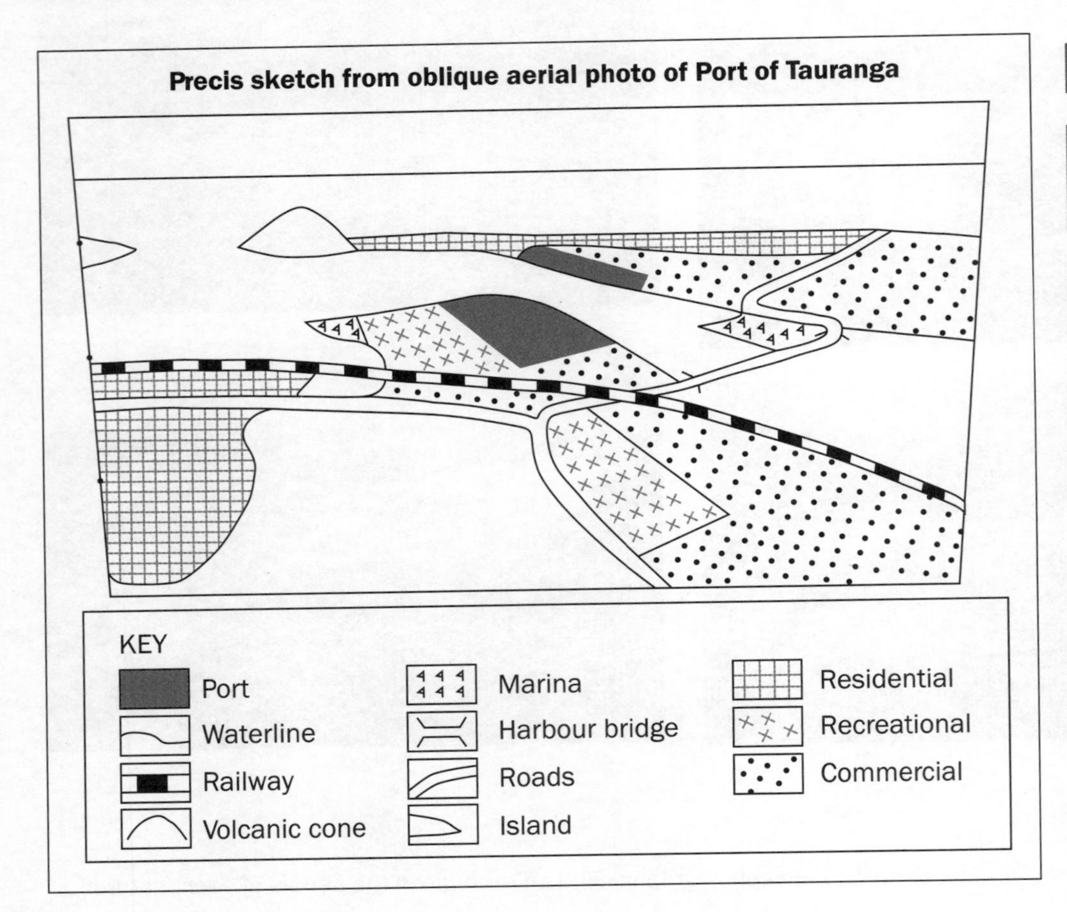

Give the sketch a title.

Draw the frame – square, rectangle, trapezium.

Draw the outline of any coastline or water.

Put dots on the frame to show halfway, then quarter-way and three-quarter-way.

Draw features as areas. For example, instead of drawing individual houses in an urban area, draw the boundary of housing.

Give the sketch a key.

 ISBN: 9780170368131

Answers

(can be removed from the centre of the book)

Unit 2

1 a 19% **b** 33%
2 a 25% **b** 66.66% **c** 20% **d** 60%
3 a 65+ **b** 0–14 **c** 15–64 Total = 100%
4 a 17.73% **b** 71.52% **c** 10.75%

Unit 3

1 a Department of Labour **b** *y*
c vertical/*y* **d** *x*
e horizontal/*x* **f** 2.5 mm
g 5.5 mm
2 a 8 **b** horizontal/*x*
c vertical/*y* **d** postgrad
e $893
3 0, 15, 25 are percents and go up the vertical axis; 2011, 2031, 2051 are dates and go along the horizontal axis.

Unit 4

1 a 0, 10, 20, 30, 40, 50, 60, 70, 80, 90
b For Graph 1, Maori population takes up all the first segment and half of the second segment. For Graph 2, it takes up one tenth of the first segment.
2 20 and 60 on percent line under graph, 'clothes' after 'Warm', 'percent' under figures below graph, 'disaster' in title, 'Source:' and 'Statistics New Zealand' together, Water before 'for three days', Torch after can opener.

Unit 5

2 a i Asian **ii** Aboriginal and other
iii White
b i Other **ii** Black African
iii Indian/Asian
iv & v Either Coloured or White

Unit 6

1 a rapid increase **b** fluctuation
c slow decrease **d** rapid decrease
e slow increase
3 a New Zealanders shopping online
b 100
c Percentage of people 18+ who shop online
d steady increase
e there is no break in scale
f there is no source given

Unit 7

1 a negative **b** none
c positive
2 a Life expectancy and literacy in countries chosen at random
b life expectancy in years
c literacy percentage **d** best fit
e six **f** right
g positive **h** increases

Unit 8

1 a temperature **b** rainfall
c months **d** mm
e °C **f** 33 °C, May
g 21 °C, December **h** 165 mm, August
i 4 mm, December

Unit 9

1 a any order — title, key, scale, frame
b direction arrow, colour
3 Cook Strait between North and South Islands. Murchison just to east of Inangahua.

Unit 10

2 a W **b** S **c** E **d** SW
e NE **f** NW
3 a north **b** west **c** north **d** west
e northeast **f** southwest

Unit 11

1 natural = volcano, river, native bush, cave, lake, sea, mountain, valley, stream, mangroves, estuary; cultural = all others
2 a blue **b** heights
c building ■ **d** lighthouse
e Rocky Point **f** lighthouse
g vegetation **h** No, No
3 a Turangi **b** 41, 47
c Tongariro
d e.g. mission house, power house, tunnel, tailrace, oxidation ponds, quarry, refuse tip, cableway, road, marae, substation, school
e e.g. bay, stream, river, swamp, island, native forest, lake
f Tokaanu tailrace canal
4 a historic Maori pa **b** marae
c beacon **d** dairy factory
e trig **f** cemetery
g airstrip **h** cool-storage sheds
i kiwifruit orchard

Unit 12

Answers on inside front cover.

Unit 13

1 a North **b** western (or northern)
c longitude **d** equator

ISBN: 9780170368131

e latitude
f eastern (or southern)
g South

2 a latitude
b equator
c longitude
d 40
e 160
f 20

Unit 14

1 a scale is shown by line — drawing 12 mm distance and marking in how many km the line represents
b by words and by representative fraction

2 line (as on the map), words (1 cm = 200 metres)

3 (distances approximate)
a 1300 m
b 1800 m
c 1200 m
d 900 m
e 800 m
f 1800 m
g 3000 m
h 2100 m

Unit 15

any order — no title, no key, no north direction, no scale, river goes over frame, frame not made with ruler

Unit 16

1 a 250 m b 200 m c 150 m d 100 m
e 50 m f 250 m

2 a valley
b depression
c ridge
d steep slope
e hill

Unit 17

1 a river
b plateau
c mountain
d plain

2 a from A to B — 0, 50, 100, 150, 200, 200, 150, 100, 50, 0
b regular shape although left side is slightly steeper

Unit 18

1 a Greek
b how many
c places
d a common feature

3 a ✓ b ✗ c ✓ d ✓
e ✗

Unit 19

1 a anticyclone
b depression
c isobar
d 1024, 1032
e 4
f hectopascal/millibar
g stationary, warm, cold

2 a fine
b slow breeze
c New Zealand
d far apart

Unit 20

1 a vertical, horizontal
b regular
c males, females

3 a 45,000
b 71,000
c 88,000
d 105,000
e 44,000

Unit 21

1 H 2 B, K 3 I 4 J
5 C, J 6 K 7 B 8 I
9 C 10 C 11 J 12 G
13 F 14 C, J 15 B 16 C
17 G 18 B, K 19 J 20 E
21 D 22 E 23 A 24 F
25 K 26 C, D, G 27 E 28 E
29 B 30 E

Unit 22

1 any order — no title, no key, individual houses drawn, wrongly shaped frame, road goes out of frame

2 rectangle frame; key should include access road, green crops, tree belts and woods, soil with tractor marks.

Unit 23

1 red = Possible effects of earthquakes; green = Damage to the land, Damage to people and property; blue = Disease from broken sewerage; yellow = all the others

2 Good effects = improved soils, volcanic peaks become tourist attractions, material left can be used for quarrying. Bad effects = soils made poor, deaths and injuries, businesses closed, loss of earnings, tourists stay away

Unit 24

2 ✓ for **f**, ✗ for **d**

3 title = Kiwifruit orcharding; central point = Attacks on kiwifruit; rays = Leafroller caterpillar, Drought, Wind, Root rot, Frost, Hail

4 title = Coal mining; central point = Environmental problems caused by coal mining; rays = Water pollution, Dust, Noise pollution, Visual pollution, Land subsiding, Air pollution

Unit 25

Data in common = 'Extreme natural event' and 'Can kill people', so these are in the middle section; rest of data about 'Volcanic eruption' is in left circle and rest of data about 'Earthquake' is in right circle

Unit 26

a to **g** any order — cows, milking machine, water troughs, electricity, accountant, farm adviser, farm bike; **h** to **k** any order — herd testing, machine maintenance, silage-making, spraying gorse; **l** to **m** any order — water pollution, profit; **n** increasing farming knowledge

Unit 27

1 sun = Evaporation, cloud = Condensation, rain = Precipitation

2 Carbon is present in all living things and in the ocean, air and rocks. It moves in a cycle, a series

 ISBN: 9780170368131

of natural processes that makes carbon in the air available to living things, which use it, and then the carbon is returned to the air.

3 a igneous, sedimentary, metamorphic
  b arrow
  c Each type of rock is formed in a different way and can be changed into another type in a cycle where processes change the rocks over time as they meet new environments.

Unit 28

1 in order – People make a decision ..., People act to move closer ..., Sustainable development becomes ..., People make sure ...

2 in order – Warnings are given ..., The tropical cyclone ..., Civil Defence ..., Crops ...

Unit 29

1 a Malcolm Evans
  b cow, pig, sheep, lamb, hen/rooster
  c waiter d hat
  e plate f clump of pasture
  g dubious h pollution
  i drench

2 water, farming, topsoil, pollution, pasture, fertilisers

Unit 30

1 a ii b i c iii

2 a Disaster Risk Reduction
  b 18 January 2015, UN, Kobe
  c population growth, urbanisation
  d official commemoration of the disaster in Japan, UN World Conference on Disaster Risk Reduction **e** 20 years after

3 a ✓ b ✓ c ✗ d ✗
  e ✓

Unit 31

2 a 4
  b high, high, low, low
  c high, falling, falling, low, rising, falling, new, better
  d low, low, low, high

3 a 2 b 4 c 1

Unit 32

1 a Distribution
  b they are about New Zealand and Antarctica only

2 a They are about the world.
  b linear
  c topic (winds and earthquakes), one has lines and one has dots for data, one has key, one has New Zealand in the centre.

Unit 33

1 a national b regional
  c global d international
  e local

2 1✗, 2✓, 3✗, 4✗, 5✗, 6✓, 7✓, 8✓, 9✗, 10✗, 11✓, 12✗, 13✓, 14✓, 15✗, 16✓, 17✓, 18✗, 19✗, 20✓

Unit 34

1 background, culture, age, sex, personality, education, religion, socio-economic group, job, what you will gain or lose

2 in order from left: B, E, F, C, D, A

Unit 35

2 a accessibility b distance
  c technology d location

3 a location b relation
  c easier, closer d movement, town, city
  e advantage f disadvantage
  g distance, length h accessibility
  i technology, places

Unit 36

eco house B, river N, stream N, pigs A and F, toolshed B, wildflowers V, cows A and F, alpacas A, fences O, gates O, pigsty B, hedges O and V, native bush N, silage pit F, haystack F, turnip crop F, water troughs O, hothouse O and F, farmer P, greenhouse O and F, horse A, tree windbreak V, orchard F, fish pond A and F, roadside stall B, woodland V, lilies V, rainwater O, glow-worm cave N, water pump O, duck pond unit N, composting A and O, aquaponic F and O, herb garden F, wetland N, olive grove V and F, heirloom varieties F, solar panel O, eco car T, barn B, hens A and F, windmill O

Unit 37

1 Include lines of boardwalk and stream, outline of mountains, location of vegetation.

2 Use boxed notes from page 72.

3 Include line of hills, location of turbines, treeline, fenceline, location of building, location of flat land.

Unit 38

1 a 1 b 5 c 7
  d 9 e 10 f 2
  g 3

2 United we stand, divided we fall.

Unit 39

a because he is the first known person to use the term

b GIS can show many different types of data on just one map by overlaying

c collecting and processing information, e.g. putting it into GIS

d see and understand patterns and relationships
e giving examples of GIS technology
f i analyse ii convert
iii overlaid iv upload

Unit 41

1 a F b O c F
d O e F f O
g F h O i F
j O
2 a F b O c F
d F e O f O
g F h F i O
j O

Unit 42

1 a Generalisation b Explanation
c Example d Diagram
2 a A b B
3 c G (Generalisation) d E (Explanation)
f E (Example) e D (Diagram)

Unit 43

a Sri Lanka b women
c picking tea leaves d lake or river
e sloping hills
f trees, tea plantations/crops
g above sea level h happy workers
i humans changed environment, cut down rainforest to plant tea crops

Unit 44

1 top left corner with N on it
2 top middle 0 to 1000 km line
3 a title, frame, key or colour 4 column
5 bar 6 horizontal
7 x 8 vertical
9 y 10 Statistics New Zealand
11 Scientists ... say it is possible to give a warning before some major earthquakes.
12 This could be enough time to shut gas lines, stop public transport and do other things to limit damage.
13 Auckland 14 vulcanologist
15 three 16 hundreds
17 20% 18 100
19 100% 20 circular
21 360 22 future
23 best fit 24 climate
25 line and column 26 mm
27 average 28 compass rose
29 NNE 30 topography
31 cultural 32 orange
33 green 34 latitudes
35 equator 36 summary
37 steep 38 uphill
39 density 40 atmosphere
41 isobars 42 no air movement
43 pyramid 44 Venn
45 trapezium 46 system
47 cycle 48 population
49 four 50 nuclear

 ISBN: 9780170368131

1 This is Troy's precis sketch from an oblique photograph. Write down five mistakes he has made.

a ______________________________

b ______________________________

c ______________________________

d ______________________________

e ______________________________

2 In the blank box, draw a precis sketch from the photograph.

Visual summaries

Visual = showing by seeing.

Summary = leaving out unnecessary material and keeping only the necessary.

Visual summary = a mind map.

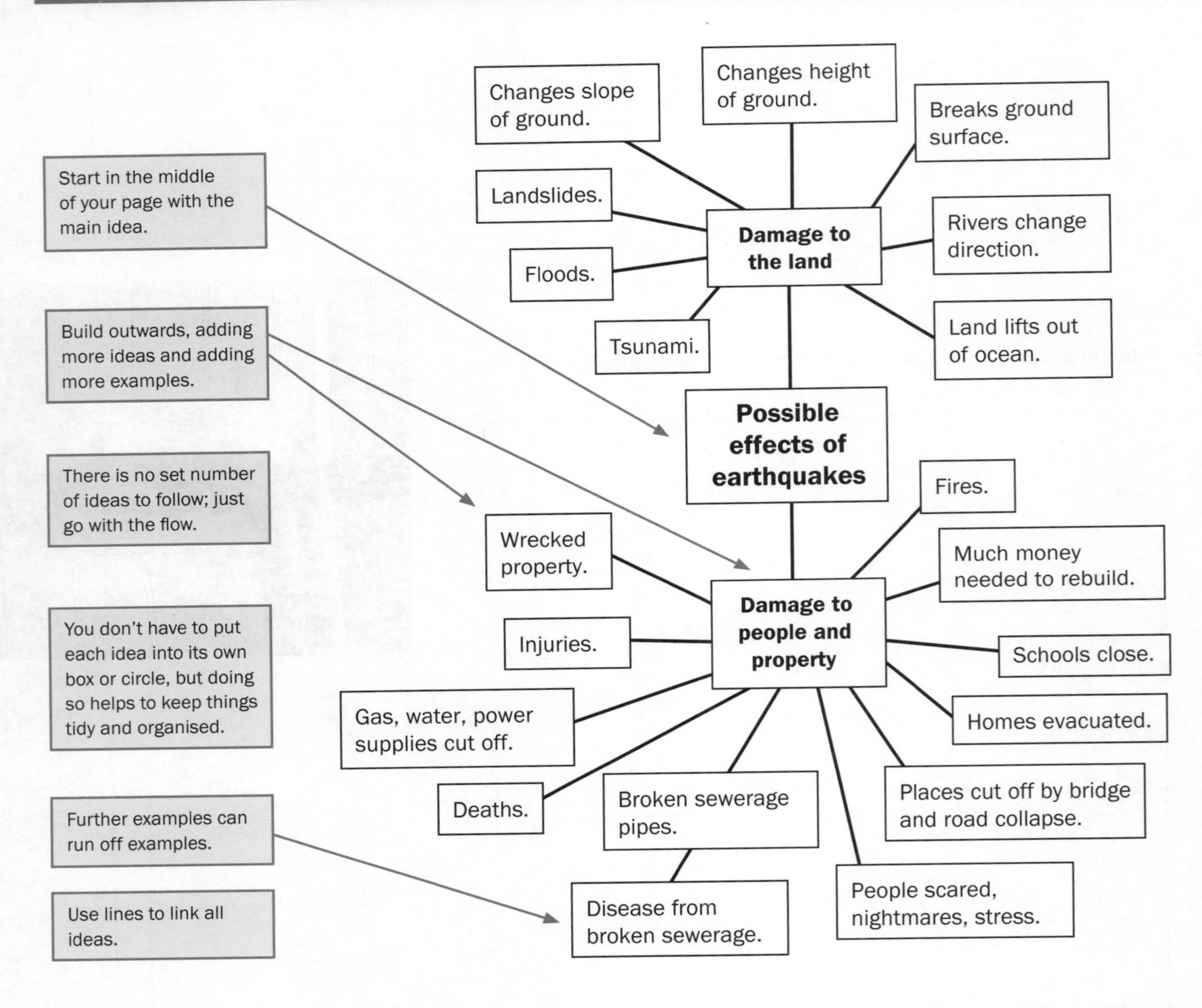

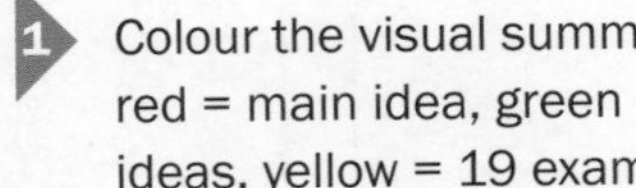

1 Colour the visual summary above as follows: red = main idea, green = two next important ideas, yellow = 19 examples of those two ideas, blue = the one example of an example.

 ISBN: 9780170368131

2 Decide whether the following ideas belong to 'Good effects' or 'Bad effects' and add them to the visual summary below: improved soils, soils made poor, deaths and injuries, businesses closed, volcanic peaks become tourist attractions, loss of earnings, material left can be used for quarrying, tourists stay away.

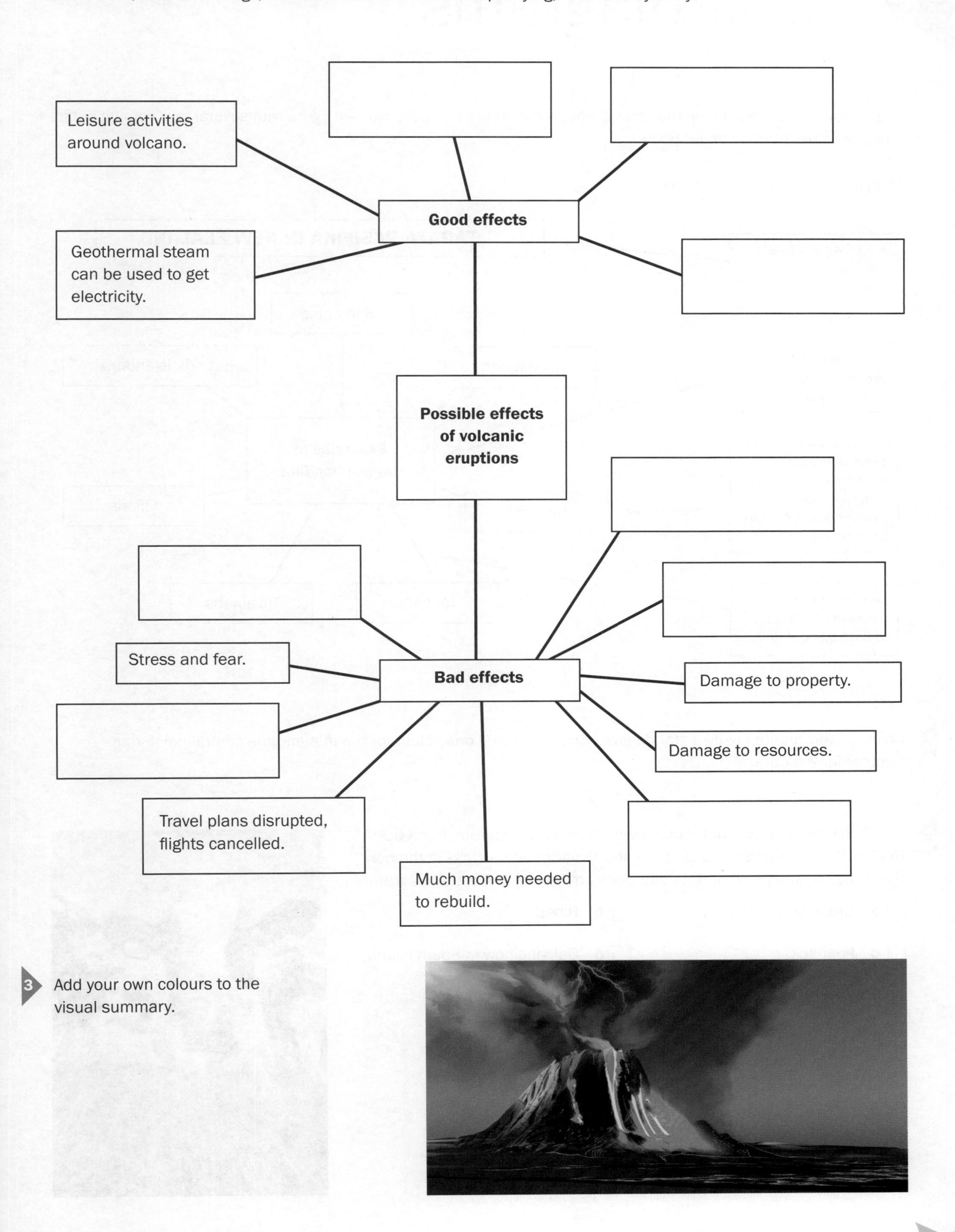

3 Add your own colours to the visual summary.

ISBN: 9780170368131

24 Star diagrams

A star diagram is made using the idea of what a star in the sky looks like – a figure with several rays coming out in a regular order from a central point.

Example of a star diagram

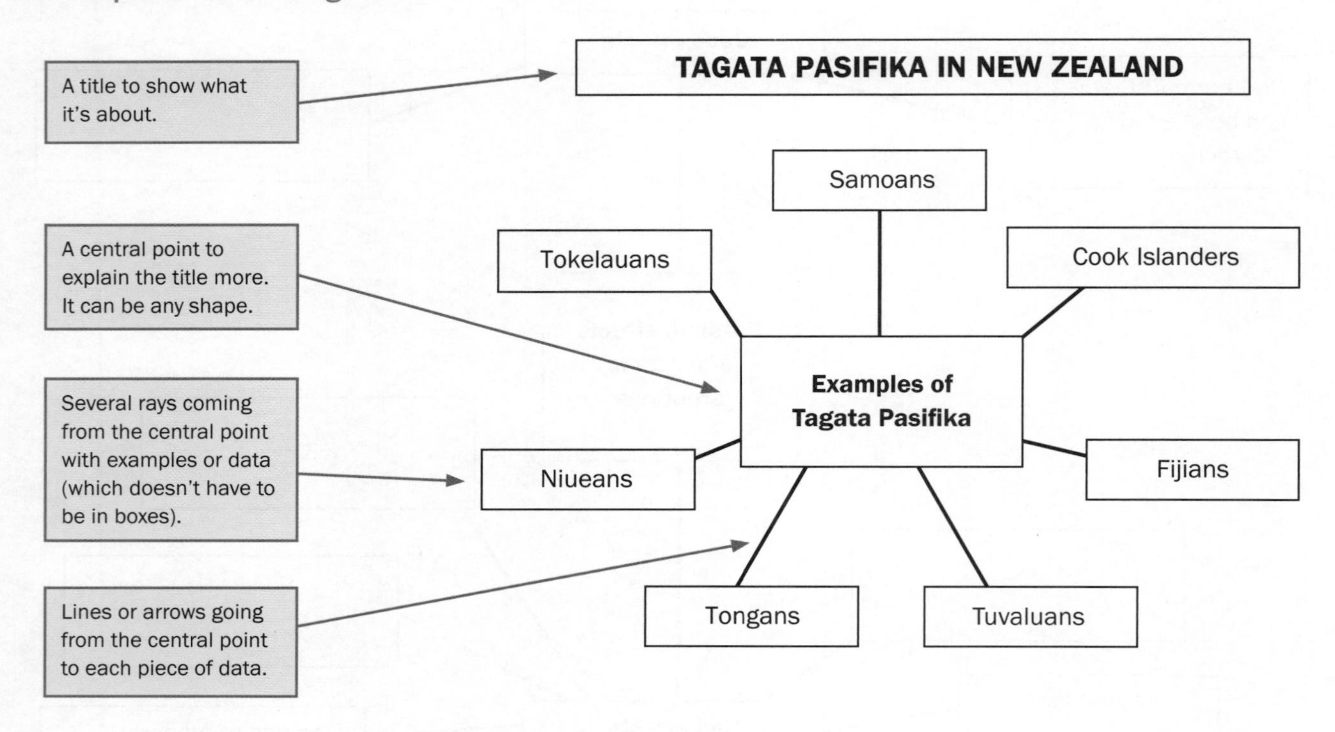

1. Colour in the Tagata Pasifika star diagram above by using one colour for the title and the central point, and another for the data.

2. The following data, except for one, belongs to a star diagram. Put a tick in the box that is beside the title of the star diagram and a cross in the box that is beside the piece of data that does not belong in the star diagram.

- ☐ **a** Grass growing slowly.
- ☐ **b** Bloat.
- ☐ **c** Poor soil.
- ☐ **d** Dairying now in South Island.
- ☐ **e** Foot rot.
- ☐ **f** Problems facing dairy farmers.
- ☐ **g** Waterlogged pasture.
- ☐ **h** Poor returns.

ISBN: 9780170368131

3 Sort out the following data and use it to fill out the star diagram.

Leaf-roller caterpillar, Drought, Attacks on kiwifruit, Wind, Kiwifruit orcharding, Root rot, Frost, Hail

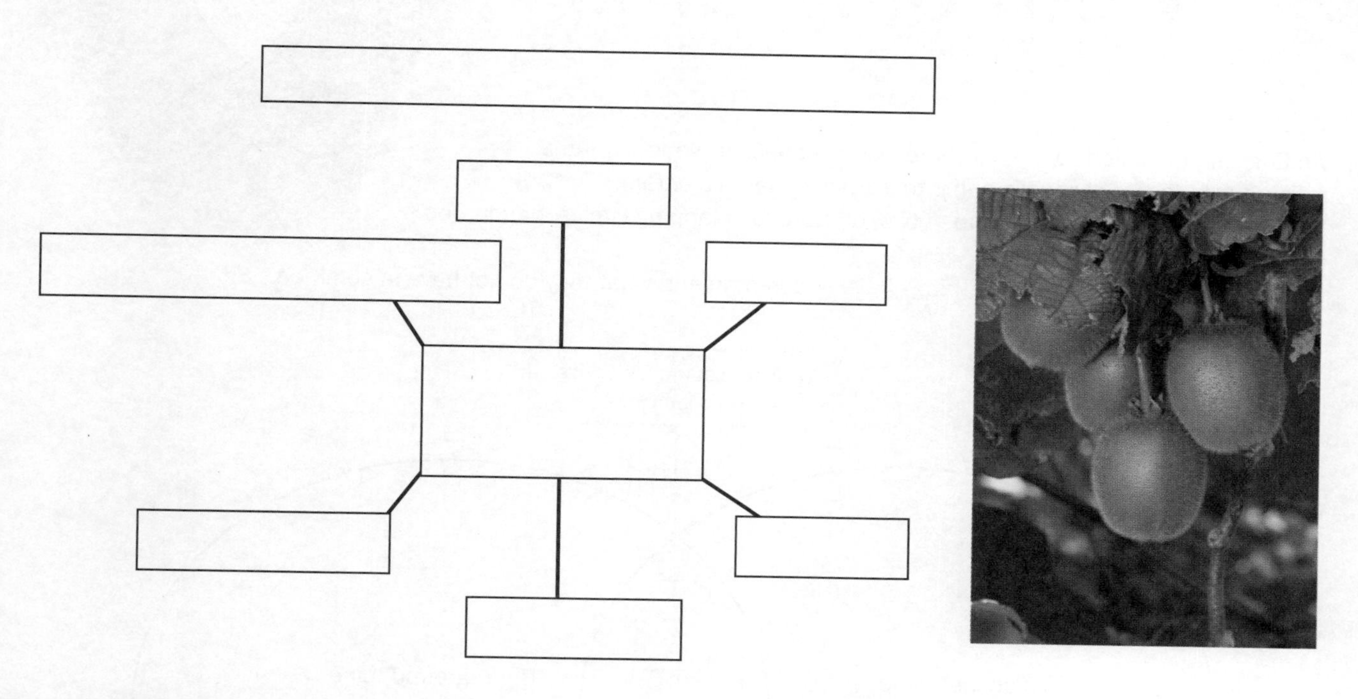

4 Sort out the following data and fill out the star diagram.

Water pollution, Dust, Environmental problems caused by coal mining, Noise pollution, Visual pollution, Land subsiding, Coal mining, Air pollution.

ISBN: 9780170368131

25 Venn diagrams

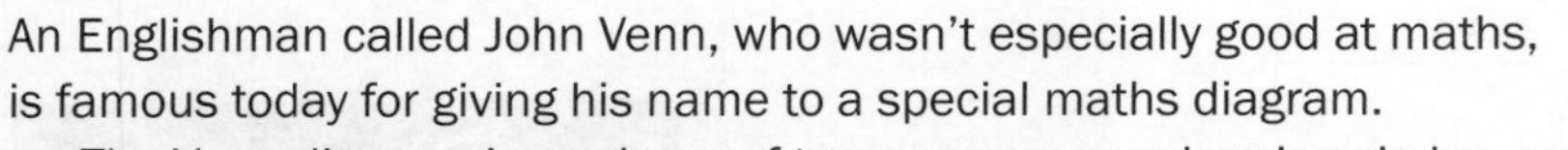

An Englishman called John Venn, who wasn't especially good at maths, is famous today for giving his name to a special maths diagram.

The Venn diagram is made up of two, or more, overlapping circles, or squares.

A Venn diagram can show what things have in common and what they do not have in common.

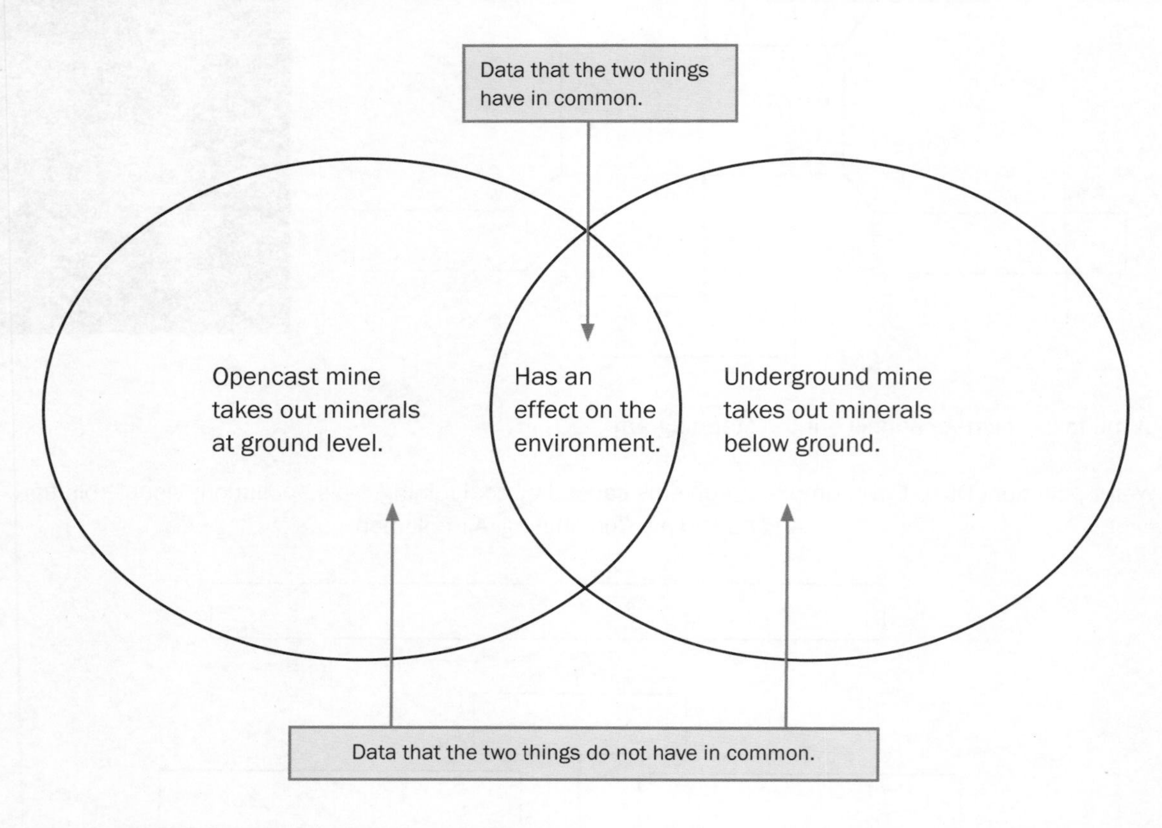

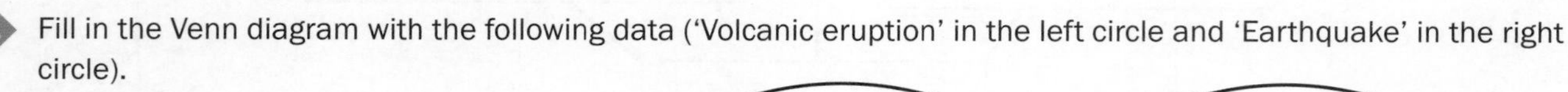

1 Fill in the Venn diagram with the following data ('Volcanic eruption' in the left circle and 'Earthquake' in the right circle).

Volcanic eruption
- Extreme natural event.
- Eruption through weakness in Earth's crust.
- Sends out hot molten rock.
- Can kill people.

Earthquake
- Sends out shock waves.
- Violent movements of Earth's crust.
- Extreme natural event.
- Can kill people.

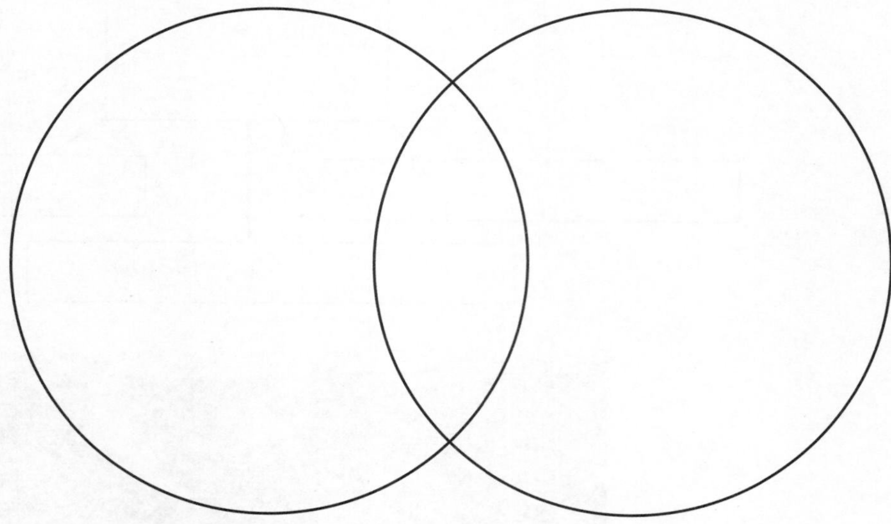

ISBN: 9780170368131

System diagrams

A system is the way something works. Your school has a system. Your favourite band has a system. Farms and mines have systems. Sustainability has a system.

Systems have four main parts. These are the four main parts of a coal mine system.

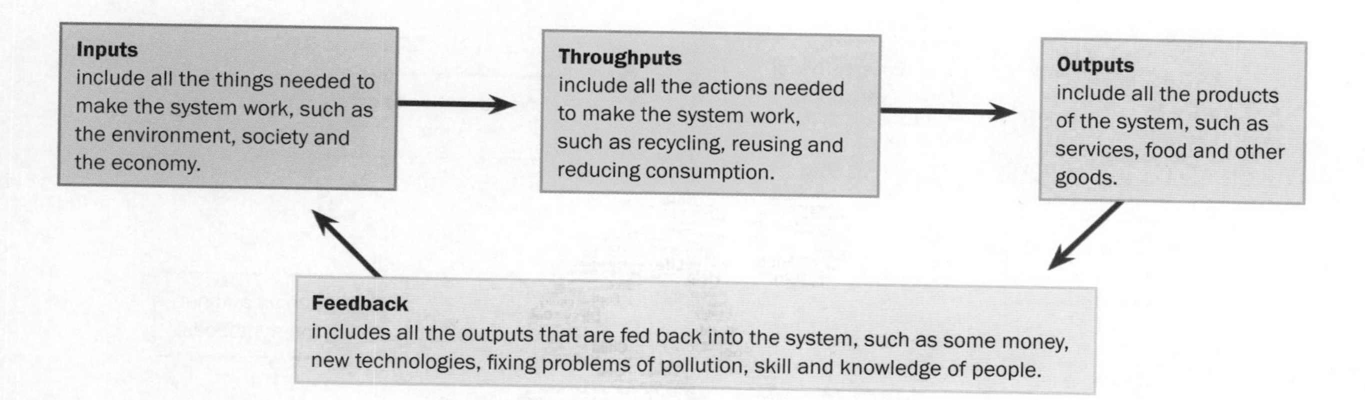

1 Write the following into the correct parts of the dairy farm system.

herd testing, cows, milking machine, machine maintenance, water troughs, electricity, water pollution, accountant, silage-making, profit, farm adviser, spraying gorse, increasing farming knowledge, farm bike.

A dairy farm system

Inputs
land, soil, rain, sun, temperatures, grass, farmer, sharemilker, herd tester, tractor, computer, milking shed, fences, effluent system.

a ______________________

b ______________________

c ______________________

d ______________________

e ______________________

f ______________________

g ______________________

Throughputs
grazing, breeding, calving, hay-making, milking, feeding out, soil testing, fertilising, sowing crops, building sheds, putting up electric fences.

h ______________________

i ______________________

j ______________________

k ______________________

Outputs
cow manure and urine, milk, female calves to replace old cows, bobby calves sold to meatworks, crops, soil erosion, country living, wear and tear on pasture.

l ______________________

m ______________________

Feedback
farm profits to build and upgrade, cow wastes as fertiliser, improving farming skills, fertilised soil.

n ______________________

ISBN: 9780170368131

27

Cycles

The word cycle comes from a Greek word meaning a circle. It is a series of events that are repeated in a regular order or time.

Around the world many people are trapped in a cycle of poverty.

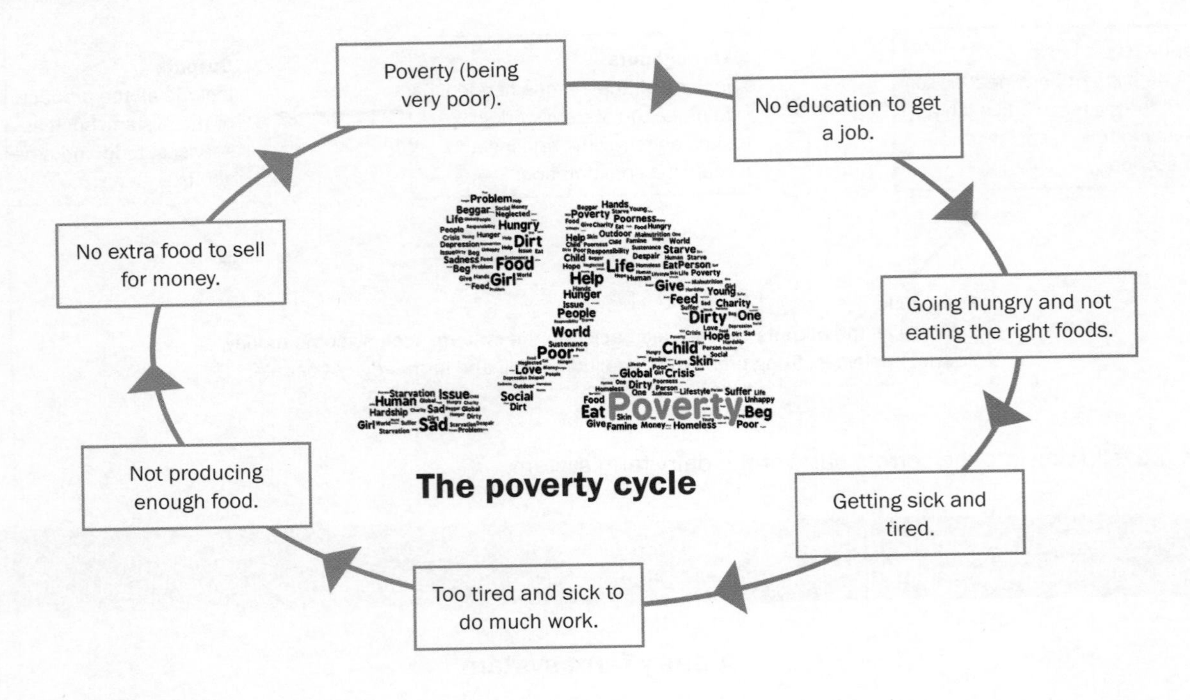

A cycle diagram of work can show events of any length of time – day, week, month, year, years.

1 The environment depends on a water cycle. Draw in the three symbols in the right place in the water cycle.

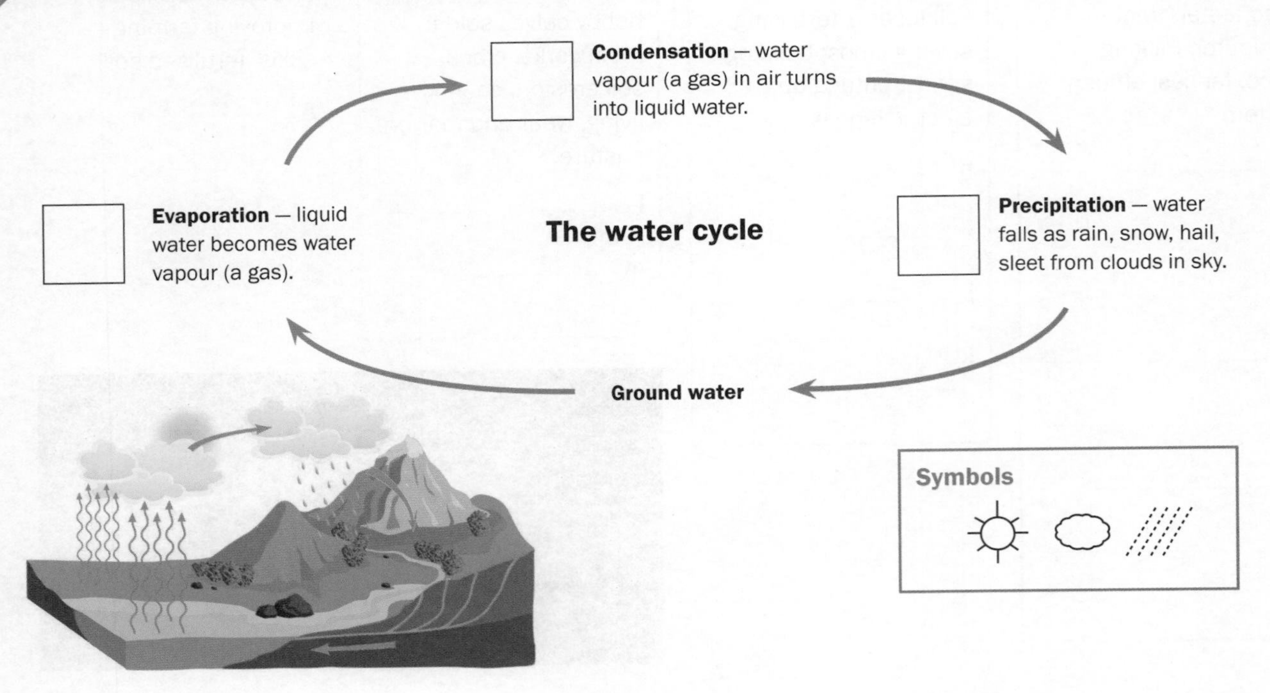

 ISBN: 9780170368131

2 Write a brief explanation of how the carbon cycle works in the environment.

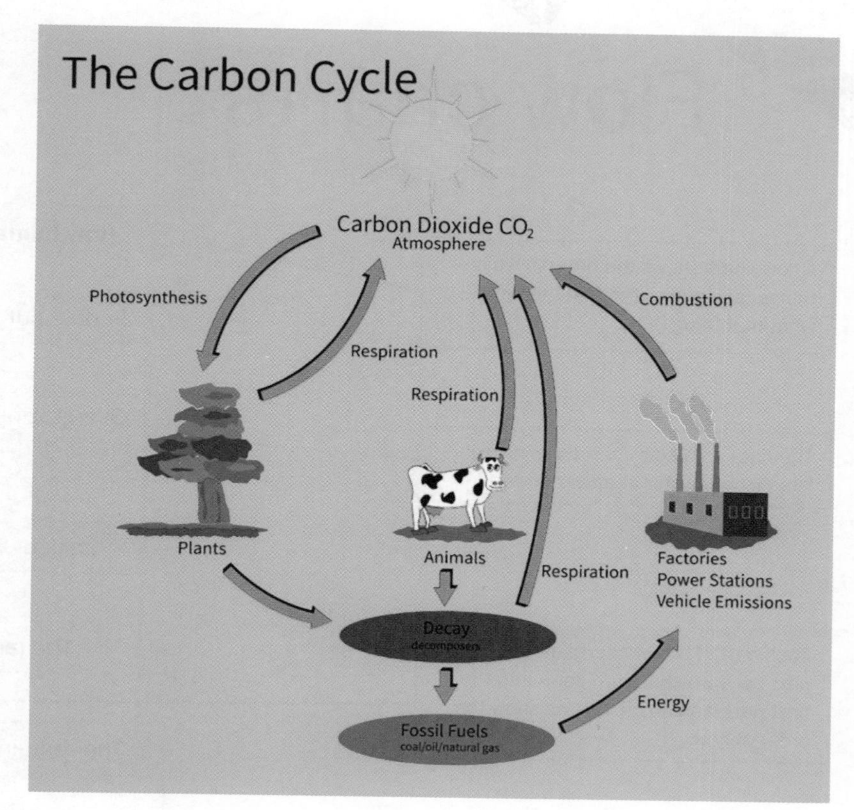

3 **a** Name the three rocks in the cycle. ______________________________

____

b Name the symbol that shows this is a cycle. ______________________________

c Say how this shows that change is part of the natural environment.

____

____

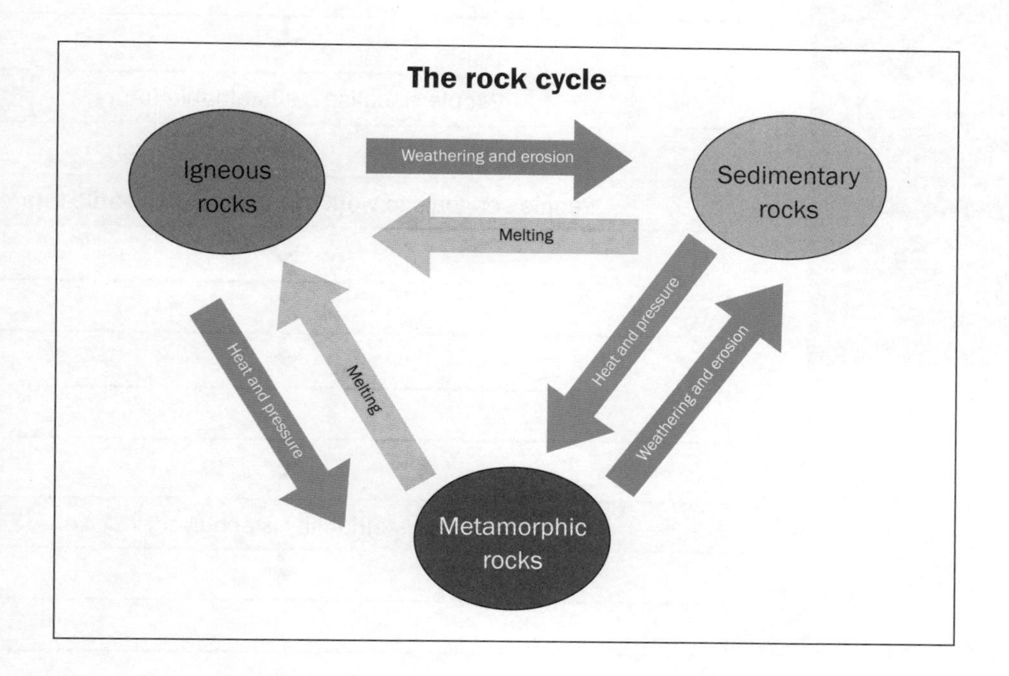

ISBN: 9780170368131

28

Flow charts

A flow chart shows the order in which things take place. The title tells you what that thing is.

The data on a flow chart flows in order like the way water in a river flows.

Each part of a flow chart flows into the next part, which in turn flows into the next part, and so on. Arrows show the order of flow.

How human activity causes desertification

People put many animals on the land to graze.

↓

Overgrazing removes the vegetation cover that binds soil to the land.

↓

Lack of vegetation cover leads to erosion.

↓

The land cannot regenerate naturally.

↓

The remaining soil is unable to recover and the land turns to desert.

1 Write the following four pieces of data into the correct boxes of the flow chart.

- People make sure sustainability is maintained.
- Sustainable development becomes more common.
- People act to move closer to goals by involving society, economy and environment.
- People make a decision to look after the environment and resources.

How sustainability can be achieved

People realise the world has limited resources.

↓

People understand they have been using resources faster than the resources can replenish themselves.

↓

____

↓

People visualise a sustainable future.

↓

People set goals to work towards a sustainable future.

↓

____

↓

____

↓

Sustainability is achieved.

↓

____

ISBN: 9780170368131

2 Write the following four pieces of data into the correct boxes of the flow chart.

- Civil defence gives help immediately.
- Crops are replanted.
- Warnings are given out to people.
- The tropical cyclone arrives.

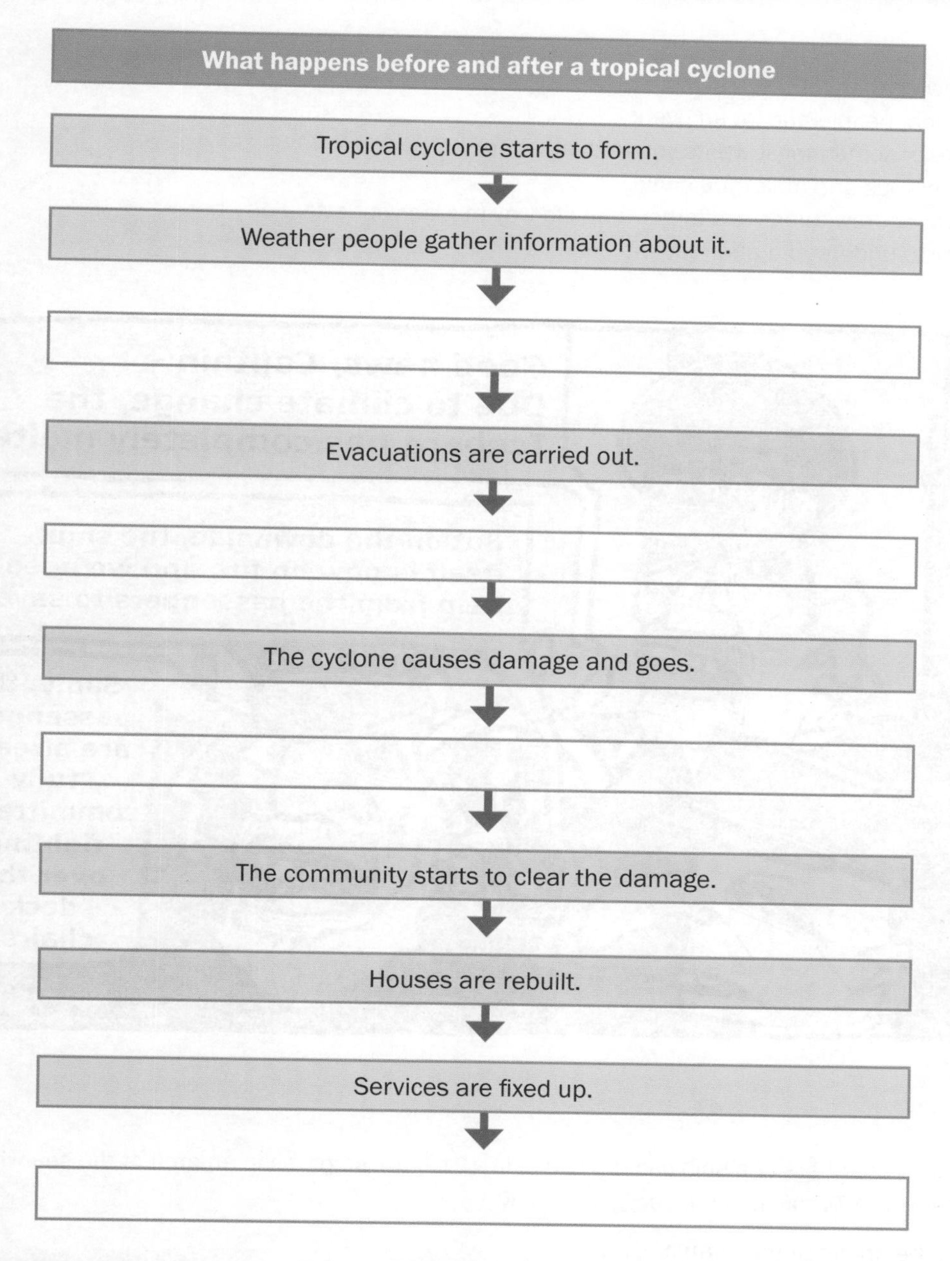

What happens before and after a tropical cyclone

Tropical cyclone starts to form.

↓

Weather people gather information about it.

↓

[ ]

↓

Evacuations are carried out.

↓

[ ]

↓

The cyclone causes damage and goes.

↓

[ ]

↓

The community starts to clear the damage.

↓

Houses are rebuilt.

↓

Services are fixed up.

↓

[ ]

ISBN: 9780170368131

29 Cartoons

A cartoon is a drawing that
- is about a person, people and/or an event
- has a message or something it wants you to think about
- aims to be humorous and make you smile
- may exaggerate or draw things in a larger-than-life way to grab your attention
- should be easy to understand and read.

Martin Doyle (1956–), 'Climate *Titanic*', 4 January 2015

The people are crew members of a ship called *Titanic*. In 1912, the ship was in an area of the sea where icebergs are usually present. The *Titanic* hit an iceberg and sank.

The setting is on the bridge of the *Titanic*.

The Captain (grey beard) is given the good news by the First Officer (has a named hat) that climate change has melted the iceberg, and the bad news by the helmsman (has ordinary unnamed cap) that the ship is on fire and the crew need the help of the passengers.

The Captain's response implies that people are concentrating on trivial matters such as deck chairs, rather than on the serious issue of climate change.

The accompanying note from the cartoonist: 'While it now seems clear that the atmosphere of the planet is warming at an alarming rate, there is an even more alarming lack of urgency from the inhabitants.'

ISBN: 9780170368131

1 Look at the Malcolm Evans cartoon and finish the statements that follow.

Malcolm Evans, *New Zealand Herald*, 28 June 2002

a The name of the cartoonist (person who drew the cartoon) is

____

b Four types of animals that have sat down to dine are ____

____

c The farmer is playing at being a ______________________

d You can tell he is a farmer because of his shirt and ______________________

e He is pretending to hold in his right hand a ______________________

f He is actually holding in his right hand a ______________________

g The word that shows the farmer doesn't know exactly what is in the pasture he is offering to diners is

____

h The fence topping is supposed to stop ______________________

i The figures in the bottom-right corner continue the farming theme by talking about

______________________ instead of wine.

2 Circle six words in the box that would be the best ones to use if you had to write about the meaning of the cartoon.

culture	water	lifestyle	farming	topsoil	gumboots
exports	pollution	table	pasture	fertilisers	pa

ISBN: 9780170368131

Interpreting resources

Interpreting = making sense of.

Resource = any material containing data and information, such as a map, graph, phone text or piece of writing.

Resource 1

Look at and think about the title.

Think about what the main idea is in the introduction to see what the story is about.

Think about the main ideas in each paragraph.

RESULTS

Think about the main idea in the concluding paragraph.

The *Rena* disaster

On 5 October 2011, the ship MV *Rena*, owned by a Greek shipping company, ran aground on the Astrolabe Reef about 12 nautical miles off the coast of Tauranga in New Zealand's Bay of Plenty. The ship had been under pressure to reach Tauranga before the ebbing tide made it unsafe to get into port.

The *Rena* carried about 1400 containers, a small number of which contained hazardous material, 1700 tonnes of heavy fuel oil and 200 tonnes of marine diesel oil. As oil began to leak into the sea and wash up on local beaches, the grounding was called New Zealand's worst-ever marine environmental disaster.

The crew was evacuated. Maritime New Zealand ordered beaches closed to the public. People were warned not to touch oil washed ashore as it was toxic, to keep windows closed if there were fumes, not to collect shellfish and not to eat food washed up on beaches from the containers.

Workers began to pump the remaining oil off the ship onto barges and to hunt for containers and spilled contents.

Several thousand people volunteered to receive training and to help clean up beaches. In white protective clothing they collected tonnes of waste, which was taken to the transfer station. Many businesses, groups and individuals provided food and drink for volunteers. Other volunteers cared for oiled animals such as seabirds and blue penguins.

Although the *Rena*'s owners apologised to the people of Tauranga, local businesses suffered, especially those directly linked with tourism and recreation, and an official report two years later said there was still contamination of Astrolabe Reef near the wreck and overall the area had not recovered. A review into the effectiveness of Maritime New Zealand's response to the disaster showed shortcomings and recommended increased resourcing to allow better responses in the future.

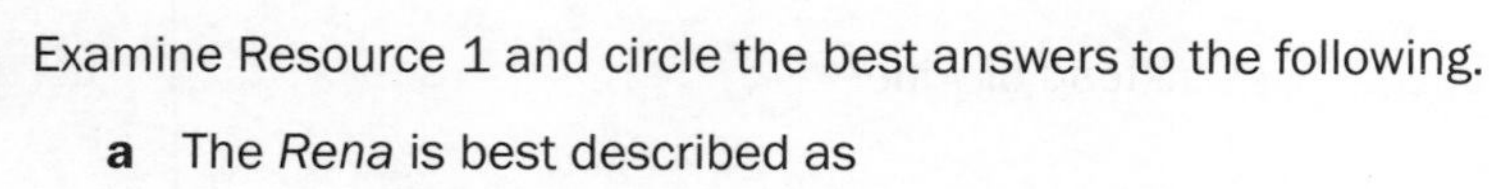

1 Examine Resource 1 and circle the best answers to the following.

a The *Rena* is best described as

i a crewed liner **ii** an oil tanker **iii** a container ship.

b Maritime New Zealand's overall response to the disaster was considered

i inadequate **ii** excellent **iii** adequate.

c The effects on the area where the disaster happened were

i few **ii** short term **iii** long term.

 ISBN: 9780170368131

Resource 2

Disaster Risk Reduction

On 18 January 2015, 20 years after 6434 lost their lives in the Kobe earthquake, the Head of the United Nations Office for Disaster Risk Reduction, Margareta Wahlström, took part in the official commemoration of the disaster in Japan, ahead of the UN World Conference on Disaster Risk Reduction, held in Japan in March, where seismic risk was to feature.

'It is important to remember distant events because a short memory is the enemy of disaster management,' Ms Wahlström said. Since Kobe, earthquakes have killed more people than any other natural hazard, with almost half of the two million deaths from major reported disasters occurring in earthquakes. Five of the most deadly earthquakes of the last 100 years had happened in the last 10 years and sent a strong message about risk and exposure in the 21st century driven by population growth and urbanisation. Ms Wahlström said proper land use and building codes were keys to reducing risk.

Look at Resource 2 and do the following.

a Underline the title.

b Highlight a date, an abbreviation, and a city.

d Underline the two factors that have increased the risk of death from earthquakes this century.

d Underline two reasons Ms Wahlström had for being in Japan.

e Underline the words that give a clue to the date of an event without actually stating the date.

Resource 3

3 Look at Resource 3 and put a tick or a cross in the boxes to show whether the statements beside them are true or false.

- ☐ **a** The words 'human' and 'people' represent society.
- ☐ **b** The three 'r' words to do with a sustainable attitude to goods appear close to each other.
- ☐ **c** Sustainability is in capitals because it is the longest word.
- ☐ **d** There are no examples of energy.
- ☐ **e** The word that is the most opposite to 'Sustainability' is 'pollution'.

ISBN: 9780170368131

31

Demographic transition models

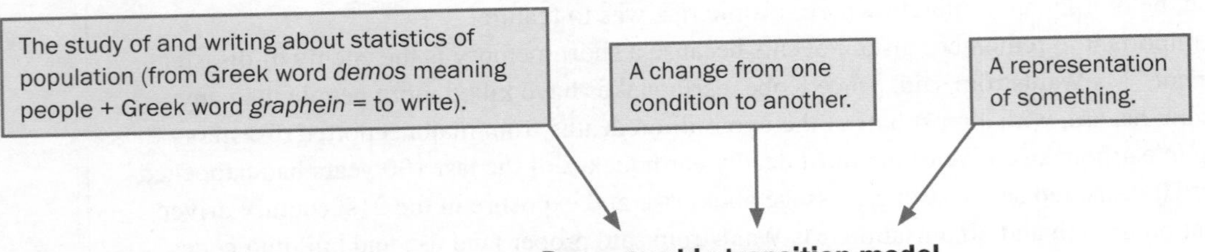

Demographic transition model
= how a country's population changes over time.

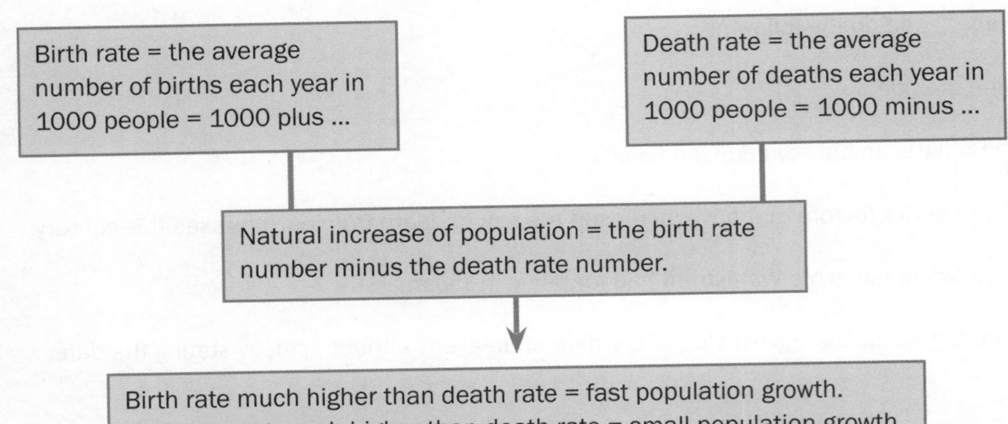

Birth rate much higher than death rate = fast population growth.
Birth rate not much higher than death rate = small population growth.
No gap between birth rate and death rate = no population growth.
Death rate higher than birth rate = loss of population.

There are usually four stages in the demographic transition model. A country will belong to one of these stages at a particular time. As the population of the country changes, the country will move into another stage on the model.

Demographic transition model			
Stage 1	**Stage 2**	**Stage 3**	**Stage 4**
high birth rate	high birth rate	falling birth rate	low birth rate
high death rate	falling death rate	low death rate	low death rate
low natural increase rate	rising natural increase rate	falling natural increase rate	low natural increase rate
low technology	new technology	better technology	high technology

ISBN: 9780170368131

1 Colour in the demographc transition model on page 60. Show your key here.

2 Fill in the gaps in the following.

a New Zealand has a low birth rate, low death rate, low natural increase rate, and high technology. This means on the model it is at Stage ____________

b In 1900, many countries in Asia were at Stage 1. This meant they had ____________ birth rates, ____________ death rates, ____________ natural increase rates, and ____________ technology.

c By 2000, many countries in Asia were at States 2 and 3. This meant their birth rates were either ____________ or ____________, their death rates were either ____________ or ____________, their natural increase rates were either ____________ or ____________, and their technologies were either ____________ or ____________

d In 2015, Japan was at Stage 4. This meant it had a ____________ birth rate, a ____________ death rate, a ____________ natural increase rate, and ____________ technology.

3 Write down the stage to which the following most likely belong.

a A population explosion because of sudden improvement in technology. ____________

b High urbanisation, technology and education, with small families. ____________

c Age-old rural society of poor farmers and large families. ____________

ISBN: 9780170368131

32 Describing patterns

Pattern = any design or system of markings.
Global = the whole world.

→ Global pattern = a pattern that is found around the world.

Global pattern can be about:

- land (such as where people live)
- air (such as where tropical cyclones happen)
- sea (such as where trading ships sail).

The easiest way to show a global pattern is to put (plot) the information on a map of the world.

→ Information that can be put on a map is called spatial data (spatial = to do with space).

Once the data has been put on a map, you can look to see what sort of pattern the data has made. Here are some main types of patterns.

Random pattern
Data has no clear design.

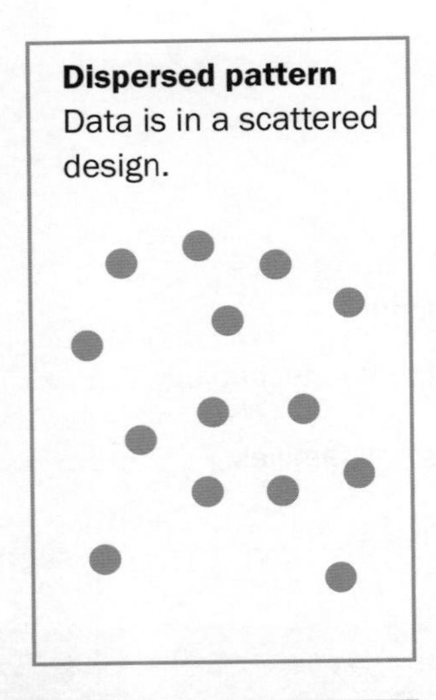

Dispersed pattern
Data is in a scattered design.

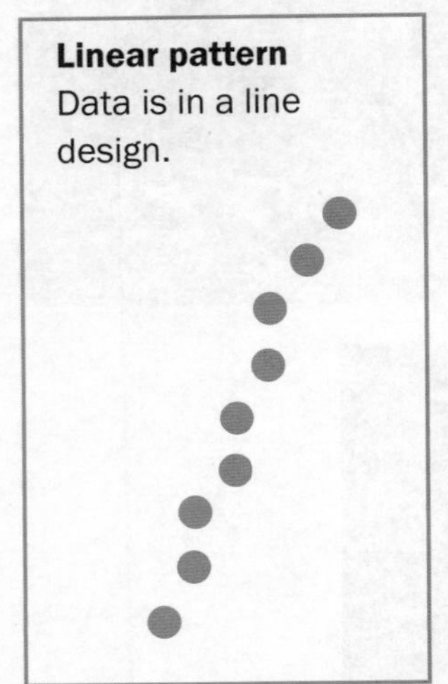

Linear pattern
Data is in a line design.

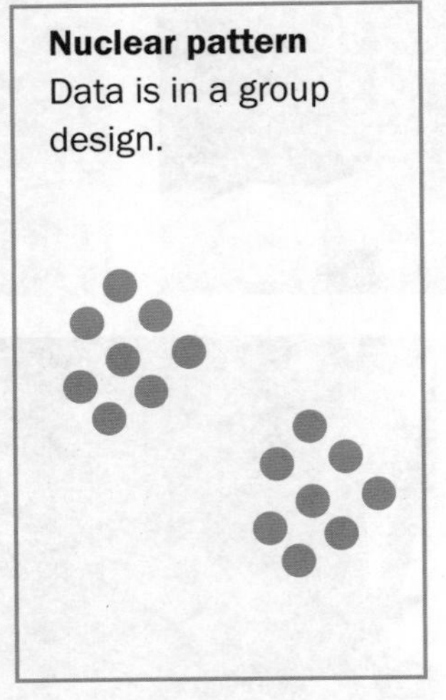

Nuclear pattern
Data is in a group design.

Distribution of coal deposits in New Zealand

North
NORTH ISLAND
Northland
Auckland
Waikato
Taranaki
Wellington
Buller
Greymouth
Christchurch
SOUTH ISLAND
Dunedin
KEY
Coal deposits

Distribution of research stations in Antarctica

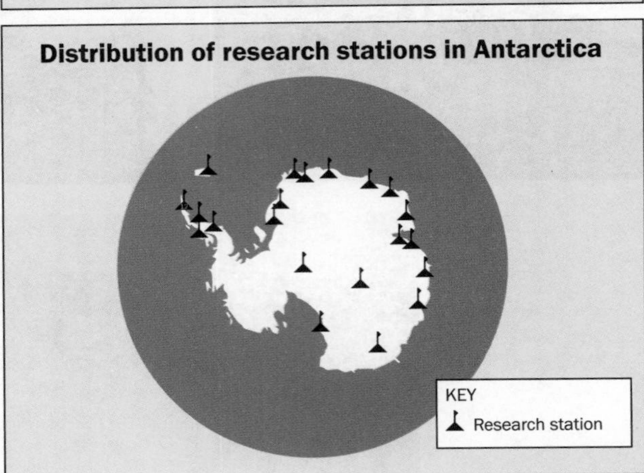

ISBN: 9780170368131

1 Fill in the gaps in the following statements about the maps on page 62.

a The word in the titles that suggests there may be patterns is ______

b They do not show global patterns because ______

______

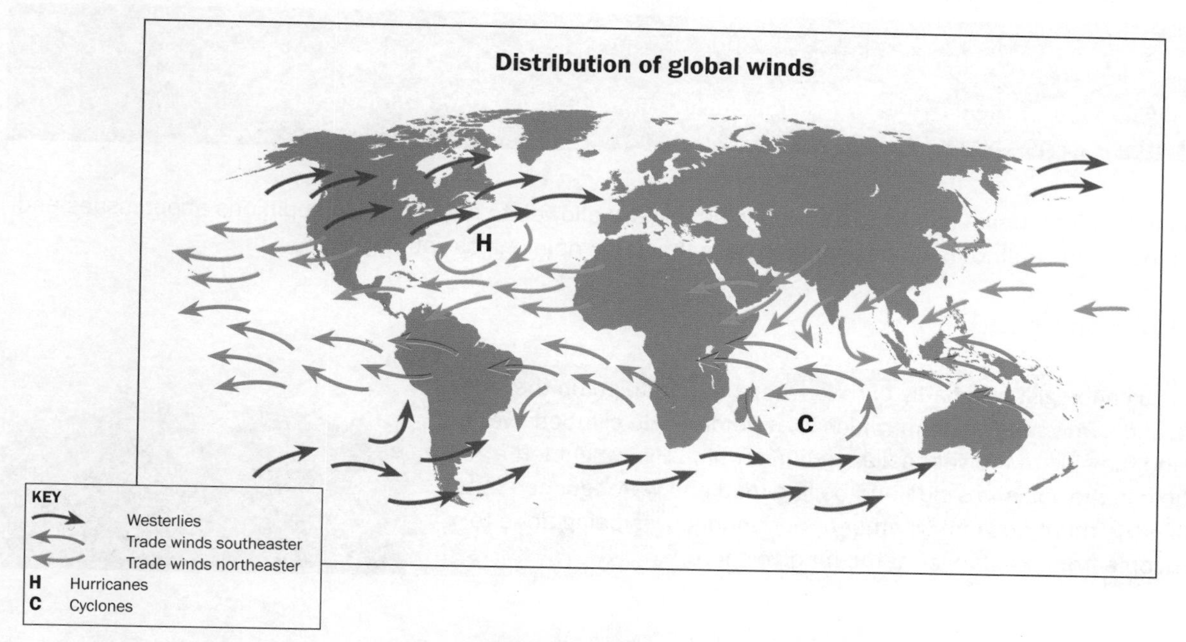

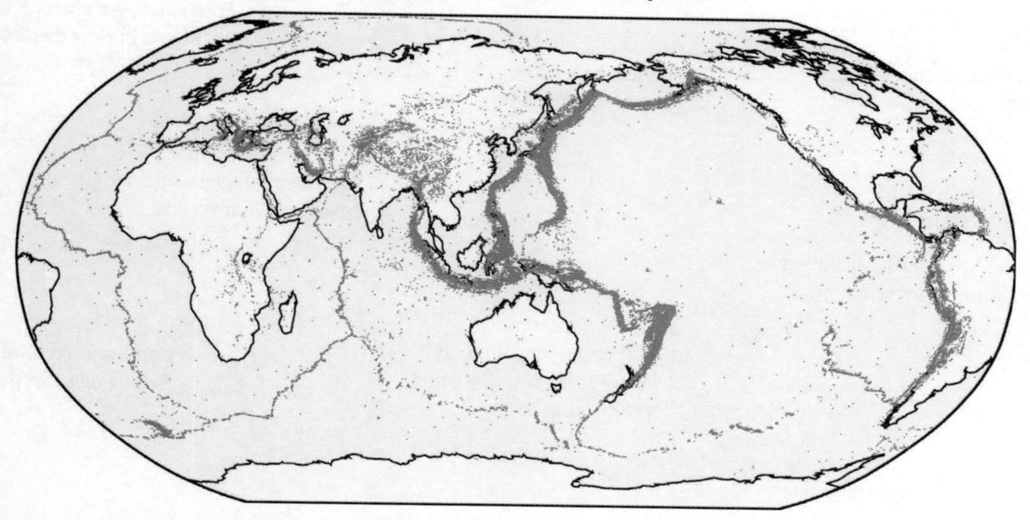

2 Look at the above two maps and state the following.

a Why they are global patterns.

______

b The most obvious pattern.

______

c Three differences between them.

______

______

ISBN: 9780170368131

33

Issues and opinions

> Issue = an event or series of events on which people have a range of opinions.
>
> Opinion = a belief or judgement that is not backed up with proof the way a fact is.

New Zealand has freedom of speech. This means people are allowed to have their own opinions about issues and say them aloud. You can't force other people to have the same opinions as you.

For example: The issue

A French vulcanologist was badly burned when he tried to go up the side of New Zealand's only mud volcano, which is at Rotorua. He climbed over the barrier and the side of the vent collapsed under him. His weight left a 70 cm hole in the volcano's side and boiling mud and water gushed out of it. The issue there became whether or not enough was being done to protect people from the danger of the mud volcano.

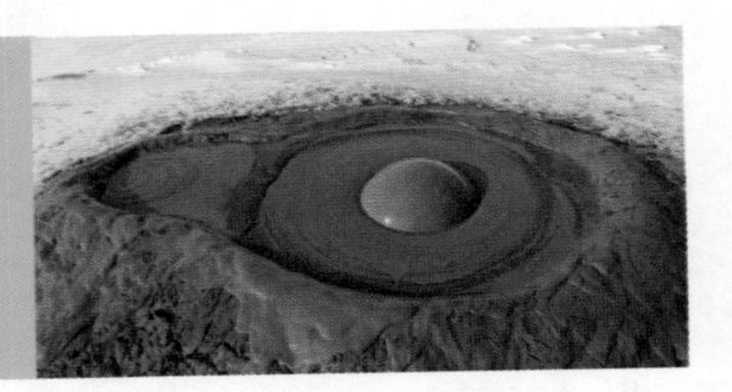

Some opinions

Foreigners are causing too many accidents in New Zealand.

Foreign visitors should bring their common sense with them.

People getting burned is not good for our tourist industry.

The vulcanologist was an idiot, so he deserved what he got.

There should be barbed wire, flashing red lights and warning signs all round the volcano.

The volcano is out to get you. People are silly going to look at it.

People should take responsibility for themselves.

Different types of issues

Contemporary = going on right now, e.g. the logging of tropical rainforest.

Historical = has been around for a long time, e.g. a 50-year-old argument over logging in a native forest.

Geographic = related to people and places, e.g. Mr X logged five kauri in a protected Coromandel forest.

Geographic issues can happen in places that are:

Local	Regional	National	International	Global
Logging in the small area you live in.	Logging in the region you live in, which may include several towns or cities.	Logging in New Zealand.	Logging around more than one country, such as the Pacific.	Logging around the world.

ISBN: 9780170368131

Example of a contemporary, geographic and national issue

Are immigrants good for New Zealand?

Some opinions

1 They take our jobs.	**2** Everyone has the right to a better life.	**3** They send money out of New Zealand to relatives.	**4** They don't understand the Treaty of Waitangi.	**5** They cause road accidents.
6 They are often young, so help balance an ageing population.	**7** They make New Zealand more interesting.	**8** They bring global talent.	**9** They increase the risk of terrorism.	**10** They increase the crime rate.
11 They bring international connections, which help trade.	**12** They keep their own cultures rather than becoming Kiwis.	**13** They make us less isolated.	**14** They pay taxes.	**15** They buy up all the houses so none is left for Kiwis.
16 They get a safe place to live in.	**17** They bring skills to the workforce.	**18** They force us to go to Australia for work.	**19** They don't understand or speak English properly.	**20** They contribute to cultural diversity.

1 State whether the following issues are mainly local, regional, national, international or global.

a The risk of New Zealand losing its clean, green image. ______________________

b Zespri's handling of the Psa vine disease in the Bay of Plenty. ______________________

c The impact of climate change on dairying around the world. ______________________

d Whaling. ______________________

e Fixing earthquake damage in Christchurch. ______________________

2 Refer to the 20 opinions above. Tick those **for** immigration to New Zealand and cross those **against**.

1	2	3	4	5	6	7	8	9	10

11	12	13	14	15	16	17	18	19	20

ISBN: 9780170368131

Value continuums

Proposed skateboard park

Values are standards on which people base their actions or opinions. They are things people think are important in life.

There are more important things in life.

Nothing's more important than skateboarding.

When people make a value judgement, they decide if something is good or bad according to their values.

Don't need a park.

Need a park.

People's background, culture, age, sex, personality, education, religion, socio-economic group, and job can make a difference to what people value.

Would rather see a temple built there.

Skating's like a way of life. You need a place to follow your way of life.

If a person is going to get something positive out of a proposed change, that person will probably be *for* the change.

Will be great for when my grandchildren come to stay.

If a person is going to get something negative out of a proposed change, that person will probably be *against* the change.

It'll disturb my peace and quiet.

1 State 10 things about yourself that will help decide your values.

 ISBN: 9780170368131

A continuum is a continuous line between two extremes to show how people feel about a particular issue.

100% For	Neutral (neither for nor against)	100% Against

In between these three points are many other positions.

For example:

The proposal: To continue uranium mining in the Kakadu region in Australia's Northern Territory.

Groups involved have different positions on a continuum.

A Aborigine residents: Mining threatens our traditional living, sacred sites and cultural treasures such as cave art. Pollution from mines threatens our health.

B Members of ERA (Energy Resources of Australia): Mining creates jobs, business and infrastructure such as roads. Nuclear energy is cleaner than fossil fuels. Mining is vital.

C Mine workers: Although there is the risk of taking uranium showers and drinking polluted water, mining creates well-paying jobs.

D Environmentalists: Despite the economic benefits, Kakadu is first and foremost a famous national park. Mining affects its ecosystems and there have already been breaches of safety standards.

E MPs (Members of Parliament): Government monitors any possible threat to the environment and legally Government owns the uranium. Mining brings huge benefits to the economy.

F Minerals Council of Australia: The uranium industry supplies fuel for near-zero emissions electricity to people around the world and for the production of medicines.

2 In the box on the continuum, put the letters belonging to the groups most likely holding the positions about Kakadu mining.

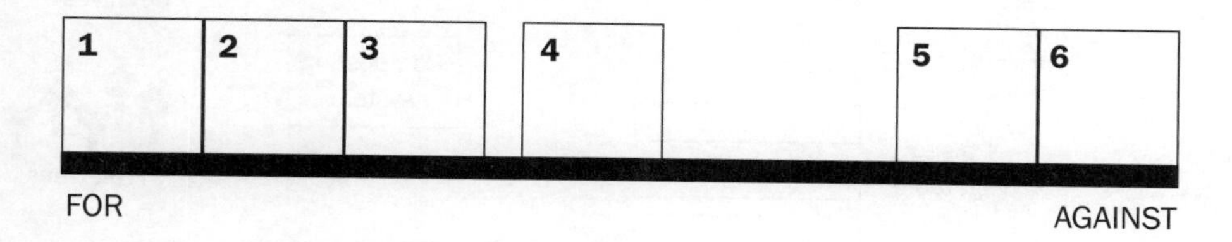

ISBN: 9780170368131

35 Applying geographic ideas

Geography is the study of Earth's surface, including its physical features, climates, vegetations, soils, population and location. Here are some geographic ideas about one of those things — location.

SOME GEOGRAPHIC IDEAS ABOUT LOCATION

- Location means where places and features are.
- You can describe location by giving directions such as latitude and longitude. Or you can say where the place or feature is in relation to other places and features.
- The closer together places are, the easier it is generally to travel between them. So there is generally more movement between the places that are closer together than the places that are far apart.
- The location of a place usually gives it advantages and disadvantages.
- Distance can be measured by length or the time it takes to get there or by how much it costs to get there.
- Accessibility is how easy it is for people, things and ideas to move around.
- Technology can make movement between places easier.

Lou has made a sketch of the sustainability project he is setting up in the corner of his father's forestry block, and used some geographic ideas about location.

LOU'S SUSTAINABILITY PROJECT

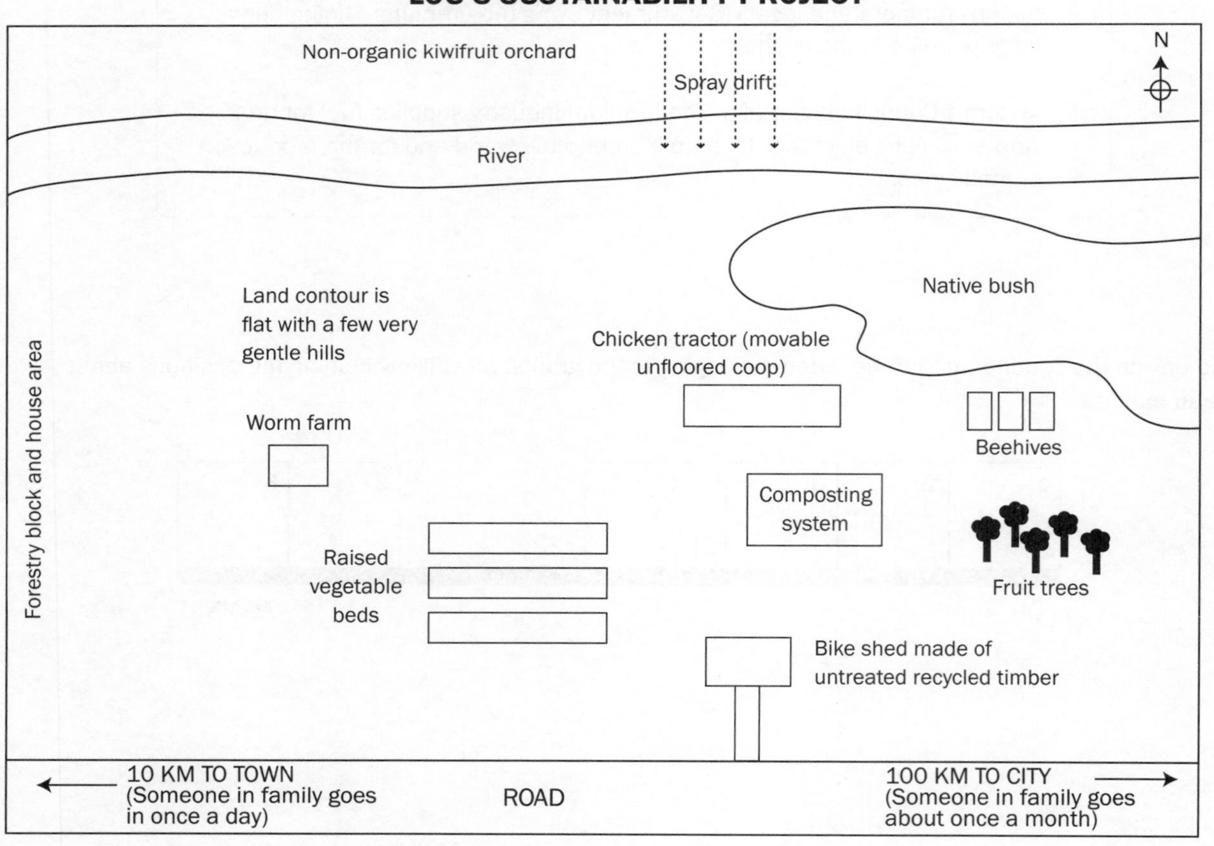

ISBN: 9780170368131

1 The words in the list below are from the text on location. Find each word in the text and circle it.

directions	location	easier	relation	features
movement	technology	closer	measured	advantages
distance	accessibility	disadvantages		

2 Give the words from the text on location that mean the following.

a How easy it is for people, things and ideas to move around. ______________________

b Can be measured by length, time or cost. ______________________

c Using machines. ______________________

d Where places are. ______________________

3 Choose words from the box to fill in the gaps in the following.

places	technology	movement	accessibility	relation
length	easier	advantage	location	distance
town	closer	disadvantage	city	

Lou has showed:

a the ______________________ of the project by using a direction arrow and road distances.

b where the project is in ______________________ to the river and the road.

c it is ______________________ to go to town than the city because the project is ______________________ to town.

d there is more ______________________ between project and ______________________ than between project and ______________________

e an ______________________ of the project's location is its nearness to the river for irrigation.

f a ______________________ is the location of a non-organic kiwifruit orchard next door.

g how ______________________ can be measured by ______________________ by using kilometres as a measurement.

h how the land contour means easy ______________________ for people and animals on the move.

i means of transport that shows how ______________________ can make movement between ______________________ easier.

ISBN: 9780170368131

36

Field work

Field = place where a particular activity happens or a particular environment is.

Field work = observation = seeing = watching = looking at = taking note of.

Field work shows you a view of the natural and cultural environment.

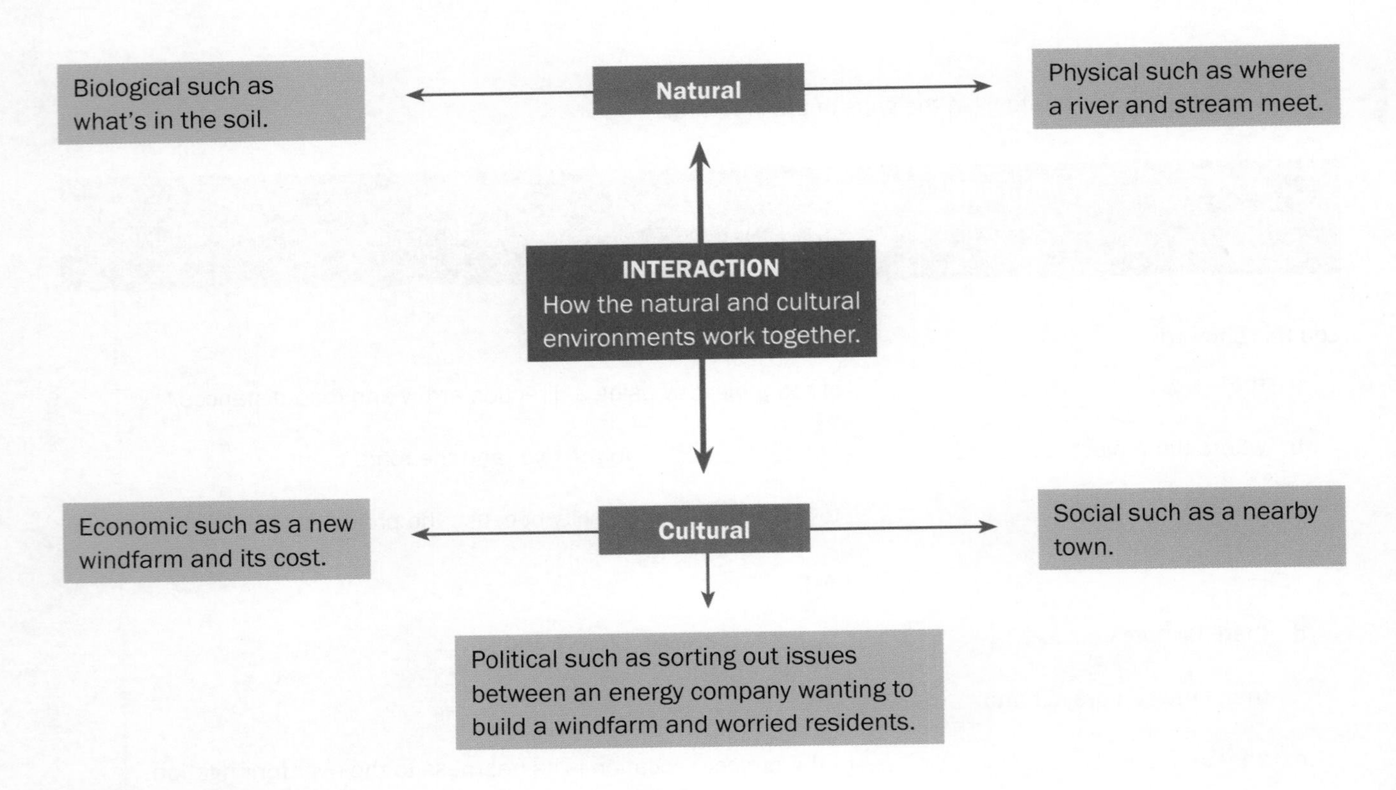

Why it is good to observe

- Helpful in emergencies, such as describing your surroundings if lost.
- Lets you see how accurate your ideas were when you thought what a particular environment would be like.
- Sharpens your powers of concentration.
- Opens your eyes to the world around you and to nature, which can make you feel better.
- Shows you how the world is made up of natural and cultural features.
- Helps you sort out how these natural and cultural features interact.
- Shows you where you fit in to this interaction between natural and cultural features.

How to do fieldwork

Nate and Claire are partners on a field trip to an eco farm. These are the steps they follow.

1. Before the trip, they list headings for features they think they might see.
2. At the eco farm, they make notes beside the headings on the list.
3. They add other features they see that are not on the list.
4. After the trip, they tidy their notes and make some general comments about what they saw.

ISBN: 9780170368131

Nate and Claire's headings

- Buildings
- Other cultural features
- Natural features
- Transport
- People
- Animals
- Fodder/food
- Vegetation

Nate and Claire's list of what they saw

- eco house of hay bales
- stream
- toolshed
- four cows
- fences of recycled wood and wire
- pigsty made of recycled materials
- native bush
- haystack
- water troughs
- farmer, wife, children
- horse
- organic orchard
- roadside stall
- patch of lilies for export
- glow-worm cave
- natural duck pond
- aquaponic vegetable unit
- wetland area
- heirloom varieties of fruit and vegetables
- barn for milking cows by hand
- freely roaming hens
- river
- two pigs and 10 piglets
- patch of wildflowers
- three alpacas
- gates
- hedges
- silage pit
- turnip crop
- hothouse
- greenhouse
- tree windbreak
- fish pond
- woodland
- rainwater collection unit
- water pump
- composting unit
- herb garden
- olive grove
- solar panel
- eco car
- windmill

1 In the boxes on Nate and Claire's list, write the first letters of the headings that the items should go under.

ISBN: 9780170368131

Field sketching

About a field sketch

- It is drawing pictures by hand unless you use digital help.
- You don't have to be a good artist, as you aren't judged on how well you draw.
- You are judged on how well you understand how features fit together.
- You can draw a scene or just a part of it.
- A photograph shows everything but a field sketch shows only those features chosen.

Why a field sketch is useful

- It shows how features interact.
- It makes you think about what you see.
- It encourages you to look more closely.
- It reduces the complicated to simple lines.
- It lets you decide which features to draw and which to leave out.
- It lets you explain what you see.
- It makes a summary of what you saw at a particular time and place.

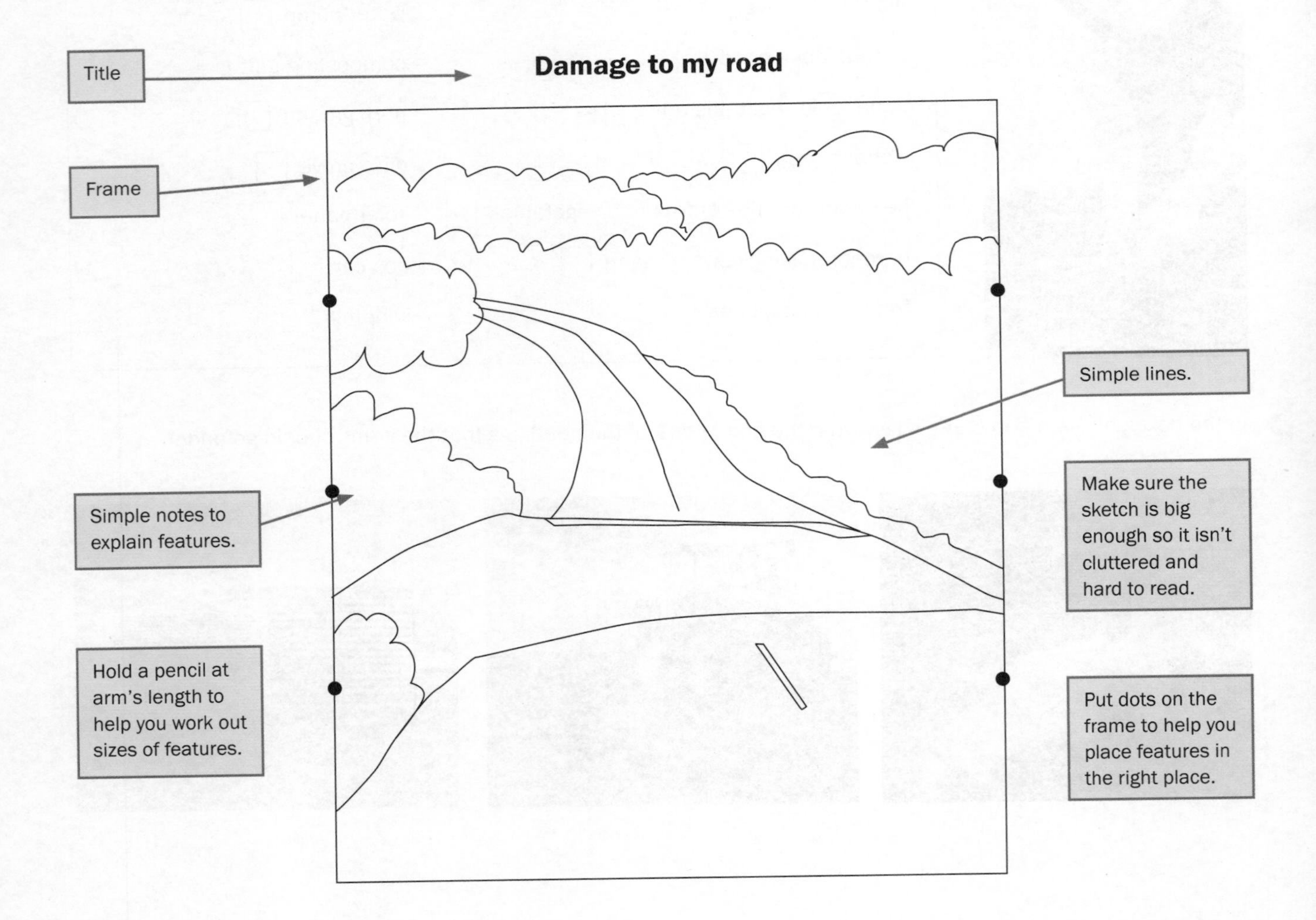

 ISBN: 9780170368131

1 You are standing on the Kea boardwalk to Aoraki/Mt Cook. In the blank box, make a field sketch to show the main features of the environment.

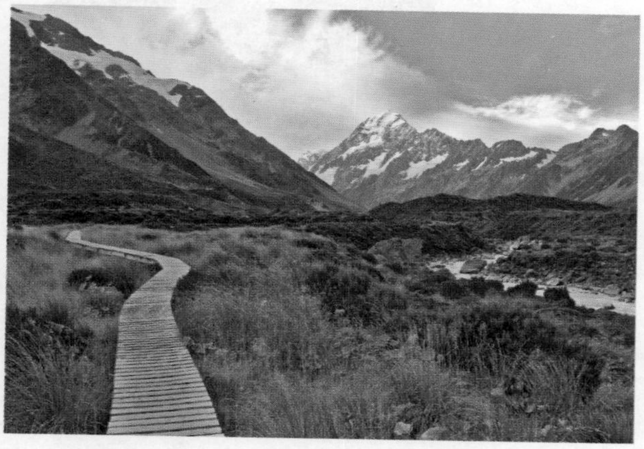

2 Write down some advice you would give the field sketcher to make sure he produces a good sketch.

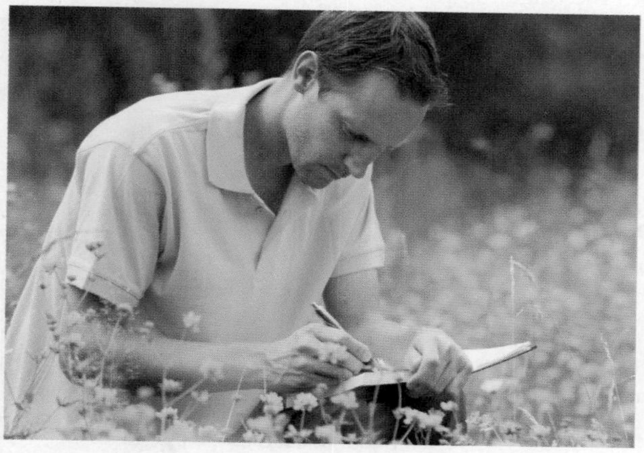

3 You are standing in a field looking at a wind farm. In the blank box, make a field sketch to show the main features of the environment.

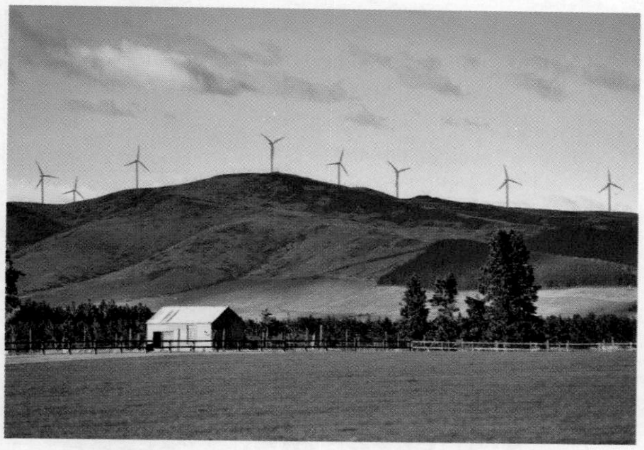

38

Group work

Why it's important to co-operate (work together)

People who study the future say future job success will depend on how well you co-operate and work well with others in groups.

Suggestions for co-operation

1. Be polite to others and realise their opinions are as important as your own.
2. Find something to be enthusiastic and positive about no matter how bad and boring things get. The more helpful you are, the better you feel.
3. Remember that what you say will often have a long-lasting effect on someone else. Find something you can praise rather than criticise.
4. Be aware there is no longer just one right way of being and doing.
5. Don't put people down just because you don't agree with them. Treat them as you want to be treated.
6. Getting angry is natural — what is important is dealing with the anger so you don't hurt others.
7. Using violence, fear, bullying or bribery to sort things out won't work for long and will cause more problems.
8. Suggest ideas. Add to others' ideas. Ask questions. Encourage others.
9. Be able to change your thinking and opinions.
10. Take your turn at playing roles such as leader, recorder, reporter.

1 Each speaker is breaking a rule about co-operation. Write the number of the broken rule in the box.

a I'm the only one in the group who knows about sustainable use of resources, so my opinion should be taken more notice of than the rest of you.

b You're dumb. Fancy not knowing what tapu means.

f This sucks. Should I care about pollution?

c I'd like to punch you. All I said was mana can apply to land as well as people.

d Nothing you say is going to change my mind. People who don't respect environmental taonga should be put in jail.

e I don't care if it is my turn to be leader. I'm not going to be and that's that. I'm sick of hearing about the environment.

g What you said is rubbish. My two-year-old brother could tell you kaitiakitanga is about looking after resources.

a	
b	
c	
d	
e	
f	
g	

2 Write the message the drawings are trying to show.

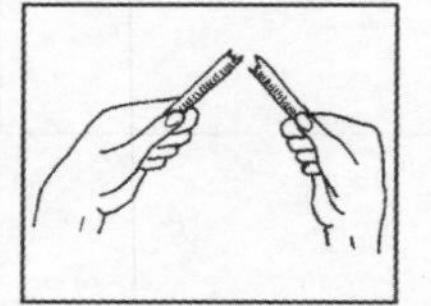 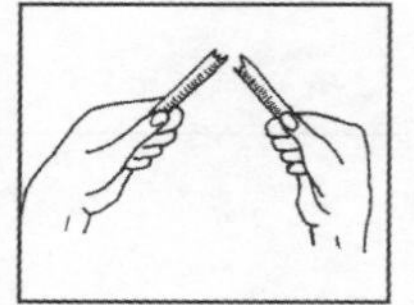 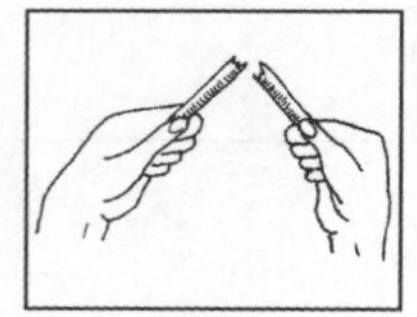 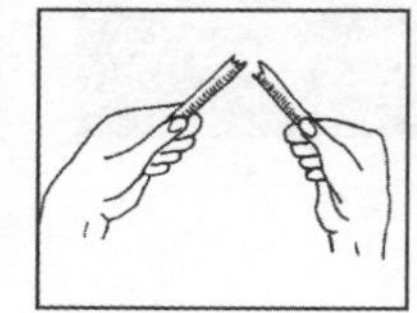 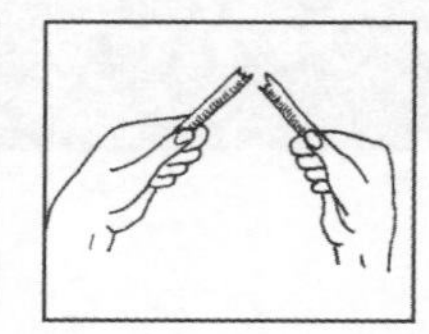 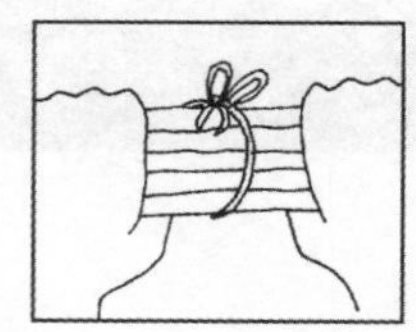

 ISBN: 9780170368131

Looking at text

Text = written words.

Looking at text = decoding (translating the written symbols) and comprehension (making sense of what is written).

Skills required = concentration, thinking, careful reading.

Read the following and answer the questions about it.

What GIS is about

GIS stands for Geographic Information System, which is a computer system designed to capture, store, manage, analyse and present geographic data.

The first known person to use the term was an English geographer called Roger Tomlinson in 1968. Today, he is acknowledged as 'the father of GIS'.

Putting information into GIS is called data capture. Information can include data about people, such as population; land, such as the location of rivers; and sites, such as where roads and storm drains are.

GIS technology can show many different kinds of data on just one map by allowing all the different types of information to be overlaid on top of one another. This lets you see and understand patterns and relationships.

You can simply upload data that is already in digital form, such as images taken from satellites and most tables, into GIS. For maps, you need to scan them or convert them into digital information.

1

a The reason Roger Tomlinson is called 'the father of GIS' is ______________________________

____

b A major difference between GIS and a map from an old book atlas is ______________________________

____

c Data capture gets its name from ______________________________

d A major benefit of GIS is that it allows you to ______________________________

e The text helps to explain the photo by ______________________________

f Find the words that mean the following.

i Examine to discover meaning. ______________________________

ii Change. ______________________________

iii Covered. ______________________________

iv Transfer from one system to another. ______________________________

ISBN: 9780170368131

40 Questionnaires and surveys

Questionnaires and surveys are sets of questions designed to get a person's opinion or gather information.

You can send them to people to fill out and send back to you either online or by post mail. Or you can ask people the questions face to face and write their answers down.

The questions are the same for everybody. This is different to an interview.

Tell people why you are doing it.

Ask first all the questions about what groups people fall into, so it is easier to sort responses into groups later.

Don't ask very personal questions. Give a range of choices.

Use yes/no answers or questions that require people to put a tick or cross in boxes.

Give an escape in case people do not fit any of the choices offered.

Keep questions simple and short. Don't use technical terms or words that people may not understand. Ask only what is important for the research. Don't ask questions that suggest the answers.

Thank the people who answer the questions for you.

Questionnaire

Reason for questionnaire/survey: a school project about the population of the town/city.

1 Male ☐ Female ☐

2 Age 0–15 ☐ 16–30 ☐ 31–45 ☐
46–60 ☐ 61–75 ☐ 76+ ☐

3 Ethnic origin
European ☐ Maori ☐ Asian ☐
Pacific Island ☐ Other ☐

4 Country of birth
NZ ☐ Australia ☐ England ☐
Scotland ☐ Ireland ☐ Netherlands ☐
Cook Islands ☐ Samoa ☐ Fiji ☐
Tokelau ☐ Tonga ☐ China ☐
Japan ☐ North Korea ☐ South Korea ☐
India ☐ USA ☐ Other ☐

5 Present address (suburb) ______

6 Length of time of residence in area
Less than 1 year ☐ 1–5 years ☐ 6–10 years ☐
More than 10 years ☐

7 Number of people living with
none ☐ 1–5 ☐ 6–10 ☐
More than 10 ☐

8 Home ownership
Own home ☐ Rent ☐

9 Main method of transport
Car ☐ Bike ☐ Walking ☐
Public transport ☐

10 Opinion on rates
Too low ☐ Reasonable ☐ Too high ☐

Thank you for your time.

ISBN: 9780170368131

Survey on recreation

Reason: to help Council prepare their next five-year plan

1 Age 0–15 ☐ 16–30 ☐ 31–45 ☐
46–60 ☐ 61–75 ☐ 76+ ☐

2 **Your most preferred recreational activities**

____

____

3 **How often do you do activities?**
Every day ☐ Weekends only ☐ 2–3 times a week ☐
4–6 times a week ☐

4 **At what time of the day do you do activities?**

____

5 **How do you get to activities?**

____

6 **What is the approximate cost of getting to activities?**

____

7 **How are the facilities that you use at the activities?**
Inadequate ☐ Adequate ☐ Crowded ☐
Not crowded ☐

8 **Which, if any, activities require adult supervision?**

____

9 **Which, if any, of your activities are unavailable in your immediate area?**

____

10 **Do you have younger family members who are interested in the same activities as you?**

____

Thank you for your time.

1 Fill out the questionnaire on page 76 yourself. Time how long it takes.

2 Fill out the survey on page 77 yourself. Time how long it takes.

3 Complete the following statements about the questionnaire and survey.

a They took about ______________ minutes in total to fill out.

b The language used was **easy/hard** to understand.

c I would feel **happy/unhappy** giving the questionnaire or survey to people.

ISBN: 9780170368131

41

Facts and opinions

Fact = a statement that is true and can be backed up with proof.

Example: Kiwifruit originated in China and were called Chinese gooseberries.

Opinion = a statement that shows what a person believes although it may not be true or able to be proved.

Example: Too much productive farmland has been converted to kiwifruit orchards and their never-ending shelter belts.

Facts and opinions about how to feed over nine million people by 2050

Facts
- The world's population is increasing.
- Runoff from farm fertilisers and manure disrupts ecosystems.
- Agriculture accelerates the loss of biodiversity because it clears areas such as forest for farms.
- Only 55 percent of the world's crop feeds people directly, while the rest is fed to livestock or turned into biofuels and industrial products.
- Farming uses much of the world's water supplies.
- Many farmers are replacing irrigation systems with methods such as subsurface drip irrigation.

Clues that show they are facts:
- Can be proved.
- No sweeping statements or generalisations such as 'People hate eating insects'.
- Few adjectives (describing words such as 'great' and 'best'.)
- Reasoned comments rather than emotional comments.

Opinions
- People consume too much meat and dairy products and too many eggs.
- The world must stop using precious food crops for biofuels.
- Consumers in the developed world need to eat less and eat leftovers more often.
- Too few people think about the food they put on their plates.
- The world has to stop increasing food production through mindless agricultural expansion.
- Governments should pass laws to make people grow their own food.

Clues that show they are opinions:
- Sweeping statements and generalisations such as 'People', which suggests everyone.
- Words such as 'must', 'need to', 'should', 'has to'.
- Adjectives such as 'precious'.
- Emotional comments such as 'mindless ... expansion'

ISBN: 9780170368131

1. Write F (for fact) or O (for opinion) in the boxes beside the following statements about whaling.

- [ ] **a** Whaling is to do with hunting whales mainly for their meat, blubber and oil.
- [ ] **b** The commercial whaling ban is the best environmental move of all time.
- [ ] **c** The International Whaling Commission put a ban on commercial whaling in 1986.
- [ ] **d** If the ban was lifted, the floodgates would open with more and more countries joining the hunters.
- [ ] **e** Japan, Norway and Iceland still kill whales.
- [ ] **f** Whaling has to stop because it is cruel and there is no humane way to kill a whale.
- [ ] **g** After the first whaling ship from America came to New Zealand waters in 1791, the seas around New Zealand became a popular place to catch whales.
- [ ] **h** Whales are greedy eaters of fish, which would mean fisheries suffer terribly if there are too many whales.
- [ ] **i** There is much controversy about present-day whaling.
- [ ] **j** It could be that pigs have been fed whale meat in some countries, which would be disgraceful.

2. Write F (for fact) or O (for opinion) in the boxes beside the following.

- [ ] **a** Environments may be natural and/or cultural.
- [ ] **b** In the battle between people and environments, people should always win.
- [ ] **c** Interaction between people and environment can cause environmental change.
- [ ] **d** Perceptions are ways of seeing an environment.
- [ ] **e** The idea of kaitiakitanga, caring for the environment, doesn't work and will never work.
- [ ] **f** Viewpoints about the environment are best left to tree-huggers.
- [ ] **g** Two examples of geographic environmental processes are erosion and desertification.
- [ ] **h** Sustainability is about people interacting with environments while making sure the environment will be available to future generations.
- [ ] **i** You should not vote for politicians who insist on changing environments.
- [ ] **j** Understanding geographic concepts won't make you a more informed citizen.

42

Writing paragraphs

A writing formula to help you organise your ideas so they make sense to other people reading them is GEED.

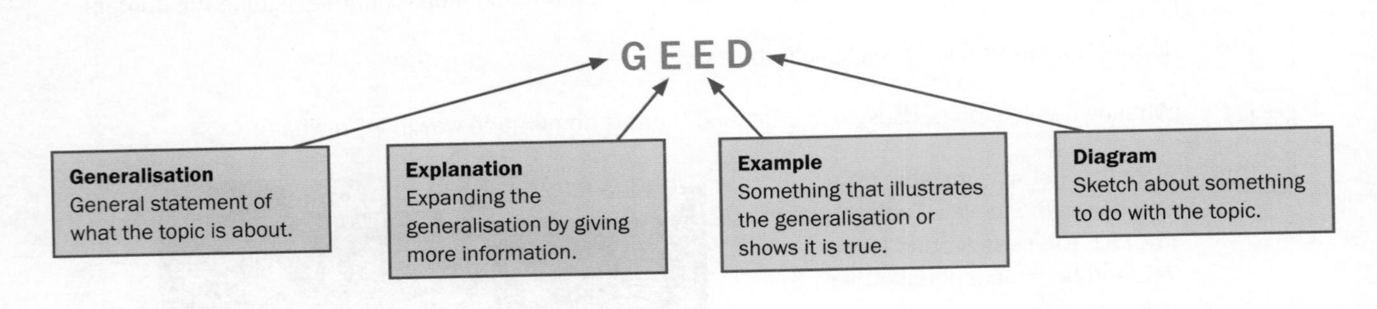

For example:

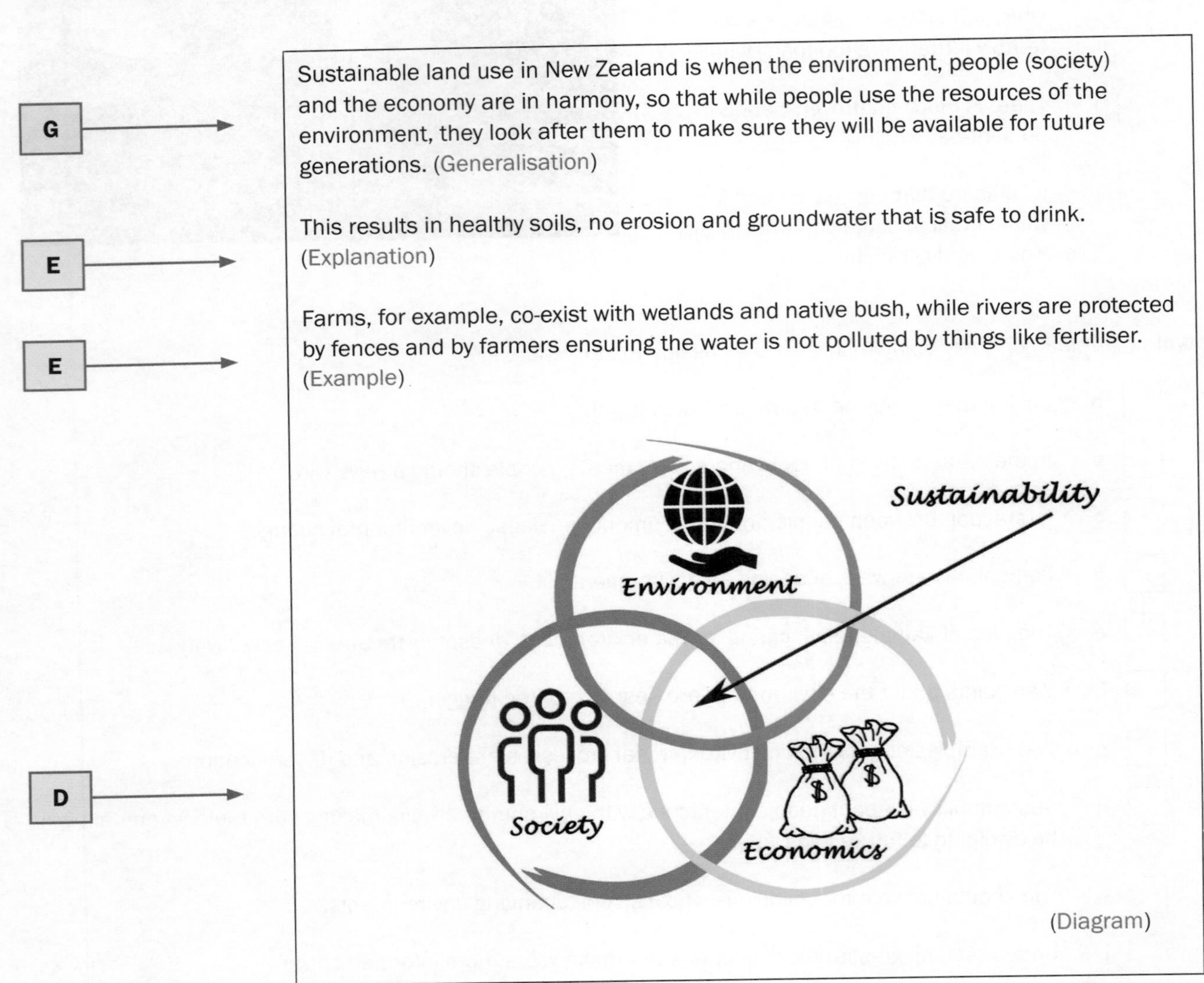

ISBN: 9780170368131

1 Sort the following parts of a paragraph into the correct order.

Parts: Example, Diagram, Generalisation, Explanation

Order: **a** ______________ **b** ______________ **c** ______________ **d** ______________

2 Look at the generalisation and circle the letters beside the best answer for the questions about it.

Generalisation = The average dairy farm income, for an average farm milking 251 cows, could drop by $40,000.

a The best example to use with the generalisation would be:

A Giant co-up Fonterra slashed about $334 million from its payout forecast for the year to next May.

B Horticulture is looking up and its export earnings are expected to pass $2 billion this year.

b The best explanation to use with the generalisation would be:

A Deer farmers have already gone from boom to bust.

B Income drop would be felt first in rural towns but would spread to the rest of the country.

3 The table has information that could help you write a paragraph about dairy production in the South Island. Choose which of the four pieces you would use by writing G, E, E, and D into four of the small boxes.

There are well over six and a half million dairy cattle in New Zealand. **a** ______	Overall, the South Island has larger farms and herds and higher production per cow and per hectare. **d** ______	
An increasing number of farms in New Zealand are converting to dairy farming. **b** ______	Otago and Canterbury **e** ______	The South Island has a significant number of large corporate farming businesses and an increasing number of irrigated farms. **f** ______
Dairy production is growing faster in the South Island, especially in Canterbury and Otago, than in the North Island. **c** ______	Dairy farming is a major contributor to the New Zealand economy. **g** ______	**Irrigation** **h** ______

43

Photographs

Look at photographs with *observation* – what you can see such as people, animals, buildings, activities, vegetation, water features, ways humans have altered the environment.

Look at photographs also with *inference* – what you can conclude about what you see such as what the climate would be like and how easy it would be to reach a place.

Taupo Bay, located in New Zealand's Far North

Observation
- Title
- Hills
- Beach
- Horseshoe-shaped bay
- Shallow seawater
- Deep seawater
- Road
- Settlement
- Gully

Inference
- Aerial view.
- Not built up like other similar bays because of limited access.
- Safe for water activities.
- Mild climate in 'the winterless north'.

Tea pickers in Sri Lanka

1. Fill out the following about the above photograph.

Observation:

a Country ______________________

b People ______________________

c Activity ______________________

d Water feature ______________________

e Contour of land ______________________

f Vegetation ______________________

Inference:

g ______________________

h ______________________

i ______________________

 ISBN: 9780170368131

Final challenge

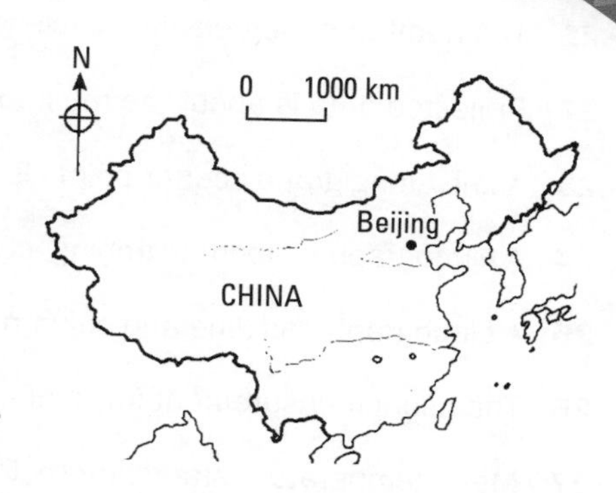

1 Circle the direction finder on the map.

2 Tick the scale on the map.

3 Three things the map needs to make it better are

____

____.

---

4 The type of graph shown here is a

______________________ graph.

5 If the bars were horizontal, it would be called a

______________________ graph.

6 The years are on the ______________________ axis.

7 Another term for this axis is ______.

8 The figures are on the ______________________ axis.

9 Another term for this axis is ______.

10 The graph's source is ______________________.

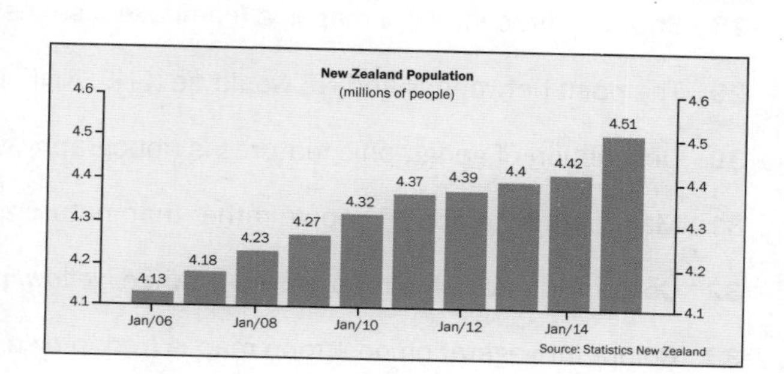

Scientists working with a network of seismic monitoring stations in Taiwan say it is possible to give a warning before some major earthquakes. 'This could be enough time to shut gas lines, stop public transport and do other things to limit damage,' said a researcher.

11 Underline the 13 words that make the key fact in the above.

12 Underline an opinion in the above.

---

13 The location mentioned is ______________________.

14 The word for an expert in eruptions is ______________________.

15 The number of possible results of a volcanic eruption is

______________________.

> A leading vulcanologist said that a volcanic eruption in Auckland would cause explosive activity to about a three-kilometre radius, pour ash over houses and workplaces, and damage transport and communications.

Circle the correct alternatives in the following.

16 Percentages means per (tens, hundreds, thousands).

17 $\frac{1}{5}$ = (20%, 24%, 30%)

18 To change a fraction to a percentage, multiply by (1, 19, 100).

ISBN: 9780170368131

19 A percentage bar graph adds up to (1%, 10%, 100%).

20 The shape of a pie graph is (square, circular, rectangular).

21 The number of degrees in a circle is (90, 180, 360).

22 Projected data is about the (past, present, future).

23 A line along graph scatter points is called (best fit, scatter, match).

24 A climograph is about (climbing, climes, climate).

25 A climograph has (line and column, scatter and line, column and scatter) graphs.

26 The usual measurement for rainfall on a graph is (cm, mm, m).

27 Mean temperature stands for (below average, average, above average) temperature.

28 Showing direction on a map is a (compass rose, news sign, path find).

29 The point between N and NE would be (ENE, NNE, NNW).

30 Description of geographic features is (topography, synergy, biology).

31 Map features made by people rather than nature are called (clone, classical, cultural) features.

32 Colour for roads on a topo map is (orange, yellow, purple).

33 Colour for vegetation on a topo map is (red, brown, green).

34 Parallel lines on a map are (latitudes, longitudes, divisions).

35 The 0 degree of latitude is the (equator, North Pole, South Pole).

36 Precis means (explanation, summary, expansion).

37 Close contour lines means land is (steep, gently sloping, flat).

38 The Vs on valley contour lines point (uphill, downhill, both up and down).

39 The word meaning 'how much' is (density, definition, diversion).

40 The layer of gases round Earth is the (barometer, atmosphere, millibar).

41 Lines on a weather map are called (forecasts, gauges, isobars).

42 A stationary front means (no air movement, fast wind, slow wind).

43 Population proportion is called a population (pyramid, populi, slope).

44 Two overlapping circles to make a third circle is a (star, Venn, system) diagram.

45 A shape of only two parallel sides is a (trapezium, rhomboid, decagon).

46 Inputs, outputs and throughputs are parts of a (system, spiral, sag).

47 Events repeated in a regular order make a (cycle, flow, resource).

48 A demographic transition model is about (relief, population, rainfall).

49 A demographic transition model usually has (four, eight, twelve) stages.

50 Data in a group design is a (linear, nuclear, dispersed) pattern.

ISBN: 9780170368131